Studies in Logic
Mathematical Logic and Foundations
Volume 80

Logics of Proofs and Justifications

Volume 70
Proceedings of the International Conference. Philosophy, Mathematics, Linguistics: Aspects of Interaction, 2012 (PhML-2012)
Oleg Prosorov, ed.

Volume 71
Fathoming Formal Logic: Volume I. Theory and Decision Procedures for Propositional Logic
Odysseus Makridis

Volume 72
Fathoming Formal Logic: Volume II. Semantics and Proof Theory for Predicate Logic
Odysseus Makridis

Volume 73
Measuring Inconsistency in Information
John Grant and Maria Vanina Mrtinez, eds.

Volume 74
Dictionary of Argumentation. An Introduction to Argumentation Studies
Christian Plantin. With a Foreword by J. Anthony Blair

Volume 75
Theory of Effective Propositional Paraconsistent Logics
Arnon Avron, Ofer Arieli and Anna Zamansky

Volume 76
Argumentation and Inference. Proceedings of the 2nd European Conference on Argumentation. Volume I
Steve Oswald and Didier Maillat, eds.

Volume 77
Argumentation and Inference. Proceedings of the 2nd European Conference on Argumentation. Volume II
Steve Oswald and Didier Maillat, eds.

Volume 68
Logic and Philosophy of Logic. Recent Trends in Latin America and Spain
Max A. Freund, Max Fernández de Castro and Marco Ruffino, eds.

Volume 79
Games Iteration Numbers. A Philosophical Introduction to Computability Theory
Luca M. Possati

Volume 80
Logics of Proofs and Justifications
Roman Kuznets and Thomas Studer

Studies in Logic Series Editor
Dov Gabbay   dov.gabbay@kcl.ac.uk

Logics of Proofs and Justifications

Roman Kuznets
and
Thomas Studer

© Individual author and College Publications, 2019
All rights reserved.

ISBN 978-1-84890-168-1

College Publications
Scientific Director: Dov Gabbay
Managing Director: Jane Spurr

http://www.collegepublications.co.uk

All rights reserved. No part of this publication may be reproduced, stored in a retrieval system or transmitted in any form, or by any means, electronic, mechanical, photocopying, recording or otherwise without prior permission, in writing, from the publisher.

Contents

Preface   vii
    Mathematical Logic Tradition vii
    Epistemic Tradition ix
    Structure of the Book xi
    What Is Missing xii
    How to Read This Book xiv
    Acknowledgments xiv

1 Modal Logic Preliminaries   **1**
    1.1 Propositional Logic 1
        1.1.1 Classical Propositional Logic 3
        1.1.2 Intuitionistic Propositional Logic 8
        1.1.3 Connectives of Variable Arity 9
    1.2 Modal Logic 11
        1.2.1 Modal Logic: Syntax 12
        1.2.2 Propositional Reasoning 14
        1.2.3 Modal Logic: Semantics 17
    1.3 Embeddings 26
    1.4 Notes 28

2 The Logic of Proofs   **29**
    2.1 Syntax 30
    2.2 Basic Properties 37
        2.2.1 Deduction Theorem 37
        2.2.2 Internalization Property 38
        2.2.3 Substitution Property 41
    2.3 Basic Modular Models 44
    2.4 Epistemic Models 51
    2.5 Notes 57

3 Relations with S4 — 61
- 3.1 Forgetful Projection 62
- 3.2 Semantic Realization 63
- 3.3 Constructive Realization 78
- 3.4 Notes 86

4 Decidability — 89
- 4.1 Post's theorem 89
- 4.2 Generated Models 92
- 4.3 Finitary models 96
- 4.4 Establishing decidability 109
- 4.5 A decidable constant specification is not enough 111
- 4.6 Notes 111

5 Complexity — 113
- 5.1 Short Introduction to Complexity Classes 113
  - 5.1.1 Class P: Deterministic Polynomial Problems ... 115
  - 5.1.2 Class NP: Nondeterministic Polynomial Problems 116
  - 5.1.3 Class coNP: Dual to NP 119
  - 5.1.4 Class PSPACE: Polynomial-Space Problems 119
  - 5.1.5 Oracle Computations and Polynomial Hierarchy . 120
- 5.2 Complexity of $\mathsf{LP_{CS}}$ 121
  - 5.2.1 Master Algorithm 122
  - 5.2.2 Best Case: Finite CS 125
  - 5.2.3 Worst Case: Irregular CS 126
  - 5.2.4 Standard Case: Almost Schematic CS 127
  - 5.2.5 Lower Bounds 133
- 5.3 Notes 134

6 Arithmetical Interpretation — 135
- 6.1 An Old Problem 135
- 6.2 Preliminaries 138
  - 6.2.1 Primitive Recursive Functions 138
  - 6.2.2 Recursive Functions 144
  - 6.2.3 Peano Arithmetic 146
- 6.3 Simple Arithmetical Interpretation 166
- 6.4 Arithmetical Interpretation 177
- 6.5 A Semantics of Proofs for Intuitionistic Logic 187
- 6.6 Notes 190

7 Self-Referentiality   193
   7.1 Self-Referential Knowledge 193
   7.2 Self-Referentiality of S4 196
   7.3 Self-Referentiality of Intuitionistic Logic 202
   7.4 Notes 203

Bibliography   205

Index   221

Preface

Justification logics are epistemic logics that feature the 'unfolding' of modality into *justification terms*. Instead of $\Box A$, which means A is known or A is believed, justification logics include formulas of the form $t{:}A$, which mean *A is justified by reason t*. One may think of traditional modal operators as *implicit* modalities and justification terms as their *explicit* counterparts. In a statement $t{:}A$, the justification term t may represent a formal mathematical proof of A or an informal reason for A.

Originally, Artemov developed the first justification logic, the Logic of Proofs, to provide a classical provability semantics for intuitionistic logic. In that approach justification terms represent proofs in a formal system like Peano arithmetic. Later justification logic was introduced into formal epistemology where justification terms can represent not only proofs but evidence in general. For instance, an agent's knowledge may be justified by direct observation or by communication with another agent. Mathematical logic and epistemology are the two main sources of justification logic, which we now discuss in more detail, see also [13].

Mathematical Logic Tradition

According to Brouwer, truth in intuitionistic logic means constructive provability, see, e.g., [137]. Based on this idea, Heyting and Kolmogorov gave an explicit (but informal) definition of intuitionistic truth that is nowadays known as Brouwer–Heyting–Kolmogorov (BHK) semantics for intuitionistic logic [76, 77, 84].

This semantics is widely accepted as the intended semantics for intuitionistic logic. However, it is purely informal and does not provide a precise definition of intuitionistic truth. Gödel [72] took the first step towards developing a rigorous proof-based interpretation of BHK semantics.

He considered the classical modal logic S4 to be a calculus describing properties of provability, that is he interpreted $\Box A$ as *A is provable*. Based on the idea that intuitionistic truth means provability, Gödel defined a translation $\mathsf{Gt}(\cdot)$ from intuitionistic logic IPC into S4. It follows from the results of Gödel [72], as well as McKinsey and Tarski [108], that $\mathsf{Gt}(\cdot)$ is indeed a correct and faithful embedding of intuitionistic logic into the modal logic S4. Hence we have an embedding of intuitionistic logic into classical logic with a provability operator.

Still the aim of defining intuitionistic logic in terms of classical provability was not reached because the connection of S4 to the usual mathematical notion of provability was not established. Even worse, interpreting $\Box A$ as *A is provable in a given formal system* T contradicts Gödel's second incompleteness theorem. Indeed,

$$\Box(\Box\bot \to \bot) \tag{1}$$

is a theorem of S4. However, if the $\Box$-modality is interpreted as the predicate of formal provability in T, then (1) expresses that the consistency of T is provable within T, which is impossible for a sufficiently strong T.

Thus we have the following situation where $X \hookrightarrow Y$ should be read as *X is interpreted in Y*:

$$\mathsf{IPC} \hookrightarrow \mathsf{S4} \hookrightarrow \ldots??? \ldots \hookrightarrow \text{classical proofs}.$$

In 1938, Gödel [73] suggested in a public lecture that using explicit proofs could help to obtain a provability interpretation of S4. Unfortunately, his work remained unpublished until 1995, by which time the idea of using explicit proofs had already been rediscovered. In a technical report of the University of Bern [20], Artemov and Straßen presented the Basic Logic of Proofs, which features formulas $t{:}F$ meaning *t is a proof of F*. Artemov [4, 6] developed this idea further and introduced the Logic of Proofs LP. He showed that S4 can be realized in LP and provided a classical provability semantics for LP.

Thus with Artemov's work, intuitionistic logic received the desired classical provability semantics:

$$\mathsf{IPC} \hookrightarrow \mathsf{S4} \hookrightarrow \mathsf{LP} \hookrightarrow \text{classical proofs}.$$

We present the embedding of S4 into LP and a classical provability interpretation for LP in Chapters 3 and 6, respectively. More details about the mathematical logic tradition of justification logic can be found in Section 6.1.

Epistemic Tradition

Besides the provability interpretation of $\Box A$ as *A is provable*, there is a very common epistemic reading of $\Box A$ as *A is known* that is usually attributed to Hintikka [78]. This modal approach to knowledge has been highly successful: it has a rich mathematical theory as well as many applications in computer science and philosophy [50, 107, 139].

Plato [42] suggested a characterization of knowledge as *justified true belief*. Epistemic modal logic, however, only works with two of Plato's three criteria for knowledge. Belief is modeled using possible worlds and an indistinguishability relation: one believes that a proposition A is true if A holds in all worlds that are considered possible. Trueness follows from the factivity axiom $\Box A \to A$, respectively from the reflexivity of the indistinguishability relation: if something is known, it must hold in the actual world. What is missing in the modal logic representation of knowledge is the justification component. Modal logic does not provide any means to express that there must be a justification for one's knowledge.

The modal logic approach to knowledge is centered around universal quantification, namely

A is known only if A holds in *all* possible worlds
that are indistinguishable from the actual one.

Justifications, on the other hand, bring in an existential quantifier, namely

A is known only if there *exists* a justification for A.

This interplay between universal and existential quantification is well-known in formal logic where we have that

there *exists* a proof of a formula A if and only if A holds in *all* models.

While mathematical proofs provide a paradigmatic example of justifications, there are many more forms of justifications that can be considered in a general epistemic setting such as direct observation, public announcements, or private communication.

Adding a general form of justifications to modal logic makes it possible to formalize and discuss many epistemic problems and puzzles. Most famously, justification logic has been used to reveal hidden assumptions in Gettier's [63] reasoning as well as to design systems that preempt Gettier-style counterexamples [9, 26, 67].

Dean and Kurokawa [47] propose a justification logic to formalize the central tenet of verificationism. This allows them to shed new light on the knowability paradox, which states that if all truths are knowable in principle, then all truths are in fact known. The paradox of the knower [81] is another classic puzzle that has been revisited from the point of view of justification logic [48].

Justification logics provide a novel, evidence based mechanism of truth-tracking, which seems to be a key ingredient for the analysis of knowledge. Artemov [9], for instance, employs this to provide a fair analysis and resolution of the Red Barn Example, see, e.g., [49]. Justification terms not only keep track of the sources of an agent's knowledge but also reflect the agent's whole reasoning process that lead to her knowledge. This important feature is exploited by Artemov and Kuznets to provide a quantitative solution to the logical omniscience problem [18].

The evidence-tracking mechanism of justification logic makes it possible to formalize justifications for an agent's knowledge. Beyond that it allows us to distinguish various reasons why something may not be known. For instance, Bucheli et al. [35] provide an analysis of the coordinated attack problem in the language of justification logic where it is possible to distinguish whether the content of a message is not known because the message has not been delivered or because its signature could not be verified.

The notion of knowledge captured by the modal logic S4 is inherently self-referential. Although this fact pops up in sequent-style proofs of certain S4-theorems, it cannot be expressed in the language of S4 directly. This, however, changes when we use justification logic. Kuznets [32] established that any embedding of S4 into LP necessarily requires self-referential justification assertions, that is assertions of the form $t{:}A(t)$ where the justification t occurs in the justified proposition $A(t)$. Self-referential assertions are not only intriguing epistemic objects, they also provide a special challenge from the semantic point of view because of the built-in vicious circle. In Chapter 7 we discuss Kuznets's result as well as the recent proof by Yu [147] that intuitionistic logic is also inherently self-referential.

Structure of the Book

Chapter 1. Modal Logic Preliminaries. The purpose of this chapter is to fix terminology and notation as well as some basic axiom systems. We introduce Hilbert-style systems for classical and intuitionistic propositional logic, which we call CL and IPC respectively. Further, we introduce syntax and semantics for the modal logic S4. We show how to construct maximal consistent sets and use them to define the canonical model for S4, which is essential for the completeness proof of S4. We finish this chapter with recalling the standard embeddings of CL into IPC and of IPC into S4.

Chapter 2. The Logic of Proofs. We begin this chapter with introducing the language as well as the axioms and rules for the Logics of Proofs LP and a family of its restricted variants LP_{CS} parameterized by a constant specification CS. We establish some basic facts about LP_{CS}: namely, the deduction theorem, the internalization property, and the substitution lemma. Then we turn to semantics. We define basic modular models for LP_{CS}, which are purely symbolic models since each justification term is interpreted as the set of formulas it justifies. Based on these symbolic models, we introduce epistemic (possible-world based) models for LP_{CS}, which are akin to Kripke models for modal logics. The connection between justification (in the sense of basic modular models) and knowledge (in the sense of Kripke frames) is provided by the principle that *justification yields belief*. Finally, we introduce fully explanatory models that additionally require every belief to have a justification.

Chapter 3. Relations with S4. Here we give a formal meaning to the claim that LP_{CS} is the justification counterpart of S4 and discuss the sufficient conditions on the constant specification CS. First, we show that S4 is the forgetful projection of LP_{CS}, meaning that the replacement of all justification terms in any LP_{CS}-theorem with $\Box$ yields a theorem of S4. More interesting but also much more difficult is the converse direction, i.e., the embedding of S4 into LP. Given any S4-theorem, it is possible to replace each occurrence of the $\Box$-modality with some justification term in such a way that the resulting formula is a theorem of LP_{CS}. This *realization theorem* is crucial for the classical provability interpretation of intuitionistic logic.

Chapter 4. Decidability. We first recall that the finite model property does not by itself yield decidability of a logic. Additionally, one needs the satisfaction relation on the finite models to be decidable. As it turns out, this is the main issue in showing decidability for LP$_{CS}$. We solve this problem by introducing the class of generated models for LP$_{CS}$, in which the interpretation of justification terms is generated by a least fixed point construction. Then, finite generated models, which we call finitary models, provide a decidable satisfaction relation. Showing soundness and completeness of LP$_{CS}$ with respect to finitary models yields decidability of LP$_{CS}$.

Chapter 5. Complexity. The computational complexity of the validity problem for the Logic of Proofs LP$_{CS}$ depends on the constant specification CS. Fortunately, it is possible to relegate this dependency to an oracle. This makes it possible to devise an oracle-based decision algorithm for any LP$_{CS}$. Studying the complexity of the oracle queries for different constant specifications finally yields upper complexity bounds for various LP$_{CS}$.

Chapter 6. Arithmetical Interpretation. It is an old question how to provide a classical provability semantics for the modal logic S4 and thus also for intuitionistic logic. In this chapter we first give more details about the mathematical logic tradition for LP$_{CS}$. Then we introduce Peano arithmetic and recall some of its basic proof-theoretic properties such as the Diagonalization Lemma. Making essential use of this lemma, we present two variants of arithmetical interpretations for LP$_{CS}$. Together with the embedding of IPC into S4 and the realization of S4 into LP$_{CS}$, this yields the desired classical provability semantics for intuitionistic logic.

Chapter 7. Self-Referentiality. In the last chapter, we study the inherent self-referentiality in the notion of knowledge that is axiomatized by S4 and LP$_{CS}$. Making use of generated models, we establish that there are theorems of S4 that cannot be realized in LP$_{CS}$ without calling for self-referential assertions of the form $t{:}A(t)$. Finally, we show that not only S4 but also intuitionistic logic is inherently self-referential.

What Is Missing

This book is mainly concerned with the Logic of Proofs, that is with the justification counterpart of the modal logic S4. This also means that we do not deal with the justification variants of other modal logics. In particular, we do not treat logics of justified belief.

Above we have mentioned some applications of justification logic that come from the tradition of mathematical logic and epistemology. Of course, many more applications have been explored, both within these two traditions and outside of them.

In the following we present a (necessarily incomplete) collection of some of the developments and topics in justification logic that are not covered in this book.

- Justification counterparts of modal logics other that S4 [1, 58, 74, 99]

- Logics that mix explicit and implicit knowledge [19, 96, 141]

- Multi-agent justification logics [8, 64, 142]

- Justification logics with common knowledge [3, 8, 35]

- Dynamic epistemic justification logics [25, 34, 36, 38, 100, 119, 120]

- First-order justification logics [23, 59, 143]

- Justification logics for reasoning under uncertainty [51, 65, 82, 83, 111, 115]

- Intuitionistic Logic of Proofs [7, 15, 46, 98, 105, 106]

- Applications to the logical omniscience problem [16, 17, 18]

- Connections to λ-calculi and type systems [2, 12, 27, 29, 117, 118, 129]

- The proof theory of justification logics [66, 89]

- Single-conclusion justification logics [86, 87, 88]

- Justification logics over other base logics (e.g. relevant, temporal, terminological) [33, 122, 131]

- The logic of knowing why [116, 140]

How to Read This Book

We cover several independent topics in the book. Hence, the book can be traversed in various ways. Figure 1 shows the dependencies among the different parts of the book. We use an arrow from A to B to express that A is a prerequisite for B.

For instance, in order to learn about self-referentiality, a reader may start with Chapters 1 and 2, then read Sections 4.1 and 4.2, and finally finish with Chapter 7. The part on classical provability semantics (Sections 6.5 and 6.6) depends on both Chapter 3 and Sections 6.1–6.4, which, however, can be read independently.

Acknowledgments

Thank you to Sergei Artemov, Melvin Fitting, and Gerhard Jäger for creating and developing the field of justification logic. Without them, justification logic would not exist as it is today. And beyond being excellent researchers, Sergei, Mel, and Gerhard are also great academic teachers and friends.

Thank you to Samuel Bucheli, Michel Marti, and Dieter Probst. Samuel helped us on the parts on realizability and decidability, Michel helped us on intuitionistic logic, and Dieter helped us on Peano Arithmetic.

Thank you to Kentaro Sato and Jan Walker, who carefully read this book at various stages when it was used as lecture notes for courses on justification logic at the University of Bern in the fall semesters of 2012, 2014, and 2016.

Thank you to Joannes Campell, Eveline Lehmann, Hamzeh Mohammadi, and Lukas Zenger for spotting several typos.

Thank you to all present and former members of the Logic and Theory Group at the Institute of Computer Science, University of Bern for a stimulating working environment and many coffee (and tea) breaks with interesting and inspiring discussions.

RK also thanks his colleagues at the Theory and Logic Group of the Institute of Logic and Computation at TU Wien who provided such a stimulating atmosphere at the later stages of the book's creation.

Thank you to Dov Gabbay for catalyzing the final spurt to the finish line.

But most importantly, we thank our best halves, Galina and Karin, for providing very important justifications for our lives.

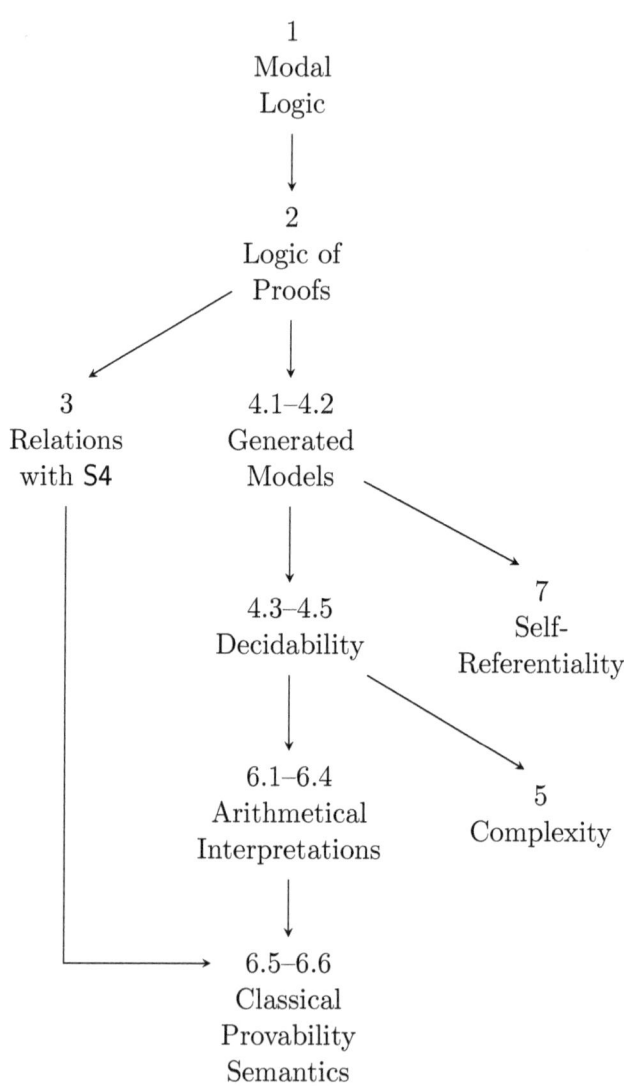

Figure 1: Dependency Diagram

Chapter 1

Modal Logic Preliminaries

The purpose of this chapter is to briefly recall syntax and semantics of the modal logic S4 and to fix the notation and terminology. Moreover we prove completeness of S4 by a canonical modal construction, which is based on maximal consistent sets. Constructions of this kind will also be used later to establish completeness of justification logics with respect to various classes of models.

An advanced reader can safely skip this chapter, using it only as reference material for notational conventions.

1.1 Propositional Logic

Propositional reasoning is the most basic level of logical reasoning. It is part of all logical systems considered in this book. In some sense, propositional reasoning is the common core of all these logical systems. However, already on this level the choice of logical rules is not unique. In this book, we consider two types of propositional reasoning: *classical* and *intuitionistic*. As the name suggests, *Classical Propositional Logic* CL is the older of the two. *Intuitionistic Propositional Logic* IPC was a reaction to certain perceived philosophical shortcomings of classical reasoning. One of the most celebrated differences is the intuitionistic rejection of the classically valid *tertium non datur* principle, i.e., of the *Law of Excluded Middle*. Given the choice between classical and intuitionistic reasoning, we take a neutral position and will not reproduce *pros* and *contras* either way. On the other hand, we have to make a clear distinction between the two reasoning modes.

But first we specify the exact meaning of the terms that are commonly used in several incompatible ways.
Natural numbers start at zero.

Definition 1.1. The *set of natural numbers* is $\mathbb{N} := \{0, 1, 2, \dots\}$.

Countable sets are always infinite.

Definition 1.2. A *countable* set is any set that has a cardinality equal to that of $\mathbb{N}$, i.e., the cardinality $\aleph_0$. A set whose cardinality is strictly less than $\aleph_0$ is *finite*. A set that is not finite is *infinite*.

Substitutions are defined using a number of notations that are too symmetric to readily suggest what is substituted for what. In order to disambiguate it, we use an asymmetric notation.

Definition 1.3. Let A be a syntactic expression, x be an atomic element that may or may not occur in A, and t be a syntactic expression of the same type as x. The result of substituting t for all free occurrences of x in A is denoted by $A[x \mapsto t]$. By $A[x \mapsto t, y \mapsto s]$ we understand the result of the simultaneous substitution of t for x and s for y in A (it is possible that x and y are atomic expressions of different types). For $\vec{x} = \langle x_1, \dots, x_n \rangle$ and $\vec{t} = \langle t_1, \dots, t_n \rangle$, by $A[\vec{x} \mapsto \vec{t}]$ we understand the result of the simultaneous substitution of t_i for x_i for each $i = 1, \dots, n$, in A. Substitutions of these types are denoted by letters σ and τ with or without sub- and/or superscripts. In this case the result of applying a substitution σ to an expression A is denoted by $A\sigma$.

Remark 1.4. Most languages we consider contain no bounding operators, making all occurrences of atomic elements free. An important exception is the first-order language of Peano Arithmetic where bound and free variables have to be distinguished.

In the definition above we do not restrict the kind of atomic elements to which substitutions apply. However, we will mostly substitute formulas for atomic propositions, justification terms for justification variables, and first-order terms for individual variables.

Axioms are formulas that are grouped into axiom schemes.

Definition 1.5. We often define logical theories by their Hilbert-style axiomatizations, i.e., by stating the axioms and inference rules. By *axioms* we mean formulas that can be derived without the use of inference

1.1 Propositional Logic

rules. However, to make the definition of logical theories more compact, axioms are usually presented by a finite number of *axiom schemes*. Axiom schemes are represented by formulas in the following sense: an axiom scheme **AxSch** represented by a formula A consists of all substitution instances of the formula A. In this case, if a formula B is a substitution instance of A, i.e., can be obtained from A by a substitution, we write $B \in \textbf{AxSch}$ and say that B is an axiom, an instance of the axiom scheme **AxSch**.

Example 1.6. Let P and Q be atomic propositions. For arbitrary formulas A and B, the formula $A \to (B \to A)$ is an instance of the axiom scheme represented by the formula $P \to (Q \to P)$. In a more abbreviated form, we can say that $A \to (B \to A)$ is an instance of the axiom scheme $P \to (Q \to P)$.

Notation 1.7. Unless stated otherwise, the axioms of a logical theory are given in the form of a set of formulas representing axiom schemes.

Remark 1.8. On the propositional level, we could have achieved the same effect by introducing the substitution rule: $\dfrac{A}{A\sigma}$. The description above does postulate the substitution rule for axioms. Unfortunately, extending this rule to all theorems would have created problems in the justification logic setting.

1.1.1 Classical Propositional Logic

Definition 1.9 (Classical propositional language). We assume a countable set $\textsf{Prop} = \{P_0, P_1, \ldots, P_n, \ldots\}$ of *atomic propositions*. Formulas of the language $\mathcal{L}_{cp}$ of classical propositional logic are defined inductively as follows:

1. $\bot$, read *falsum*, is a formula of $\mathcal{L}_{cp}$;

2. each atomic proposition $P_i \in \textsf{Prop}$ is a formula of $\mathcal{L}_{cp}$;

3. if A and B are formulas of $\mathcal{L}_{cp}$, then so is $(A \to B)$, read *if A, then B or A implies B*.

We use letters P, P', P'', etc., as well as Q with or without sub- and/or superscripts, to denote atomic propositions. Letters A, B, C, D, E, F, G, and H, with or without sub- and/or superscripts, are used to denote formulas (of the current language). We do not specify this language explicitly whenever it can be inferred from the context.

As usual, we define

- $\neg A := (A \to \bot)$,
- $\top := \neg\bot$,
- $(A \vee B) := (\neg A \to B)$,
- $(A \wedge B) := \neg(\neg A \vee \neg B)$, and
- $(A \leftrightarrow B) := \bigl((A \to B) \wedge (B \to A)\bigr)$.

Negation $\neg$ binds stronger than all binary connectives. Conjunction $\wedge$ and disjunction $\vee$ bind stronger than implication $\to$ and equivalence $\leftrightarrow$. Implication binds stronger than equivalence. The binding strength of disjunction and conjunction is the same. Binary connectives of the same strength are left-associative. Parentheses can be omitted if they can be inferred from these parsing rules. For instance,

$$\neg A \leftrightarrow B \to C \vee D \wedge E \quad \text{means} \quad \Bigl((\neg A) \leftrightarrow \bigl(B \to ((C \vee D) \wedge E)\bigr)\Bigr).$$

We define subformulas of an $\mathcal{L}_{\text{cp}}$-formula as usual.

Definition 1.10 (Subformula). The set of subformulas $\mathsf{sub}(F)$ of an $\mathcal{L}_{\text{cp}}$-formula F is inductively defined by:

1. $\mathsf{sub}(\bot) := \{\bot\}$,
2. $\mathsf{sub}(P_i) := \{P_i\}$ for each $P_i \in \mathsf{Prop}$,
3. $\mathsf{sub}(A \to B) := \{A \to B\} \cup \mathsf{sub}(A) \cup \mathsf{sub}(B)$.

The *classical propositional logic* CL is described by a Hilbert-style system with axiom schemes **PAx1**, **PAx2**, and **LDN** (the *law of double negation*) and the inference rule *modus ponens* (**MP**) (see Figure 1.1).

We write $\mathsf{CL} \vdash A$ if a formula A is derivable in this Hilbert-style system. Traditionally, we use the same notation for a logic, its Hilbert-style system, and the set of its theorems. For instance, to say that $\mathsf{CL} \vdash A$, i.e., that A is derivable in the Hilbert-style system for CL, is equivalent to saying that $A \in \mathsf{CL}$, i.e., that A belong to the set of theorems of CL, and is equivalent to saying that A is a classical propositional tautology, i.e., a validity of CL.

1.1 Propositional Logic

PAx1 $P \to (Q \to P)$

PAx2 $(P \to (Q_1 \to Q_2)) \to ((P \to Q_1) \to (P \to Q_2))$

LDN $((P \to \bot) \to \bot) \to P$

Inference rule *modus ponens* $\dfrac{A \quad A \to B}{B}$ (**MP**).

Figure 1.1: The axiom schemes and inference rule of the Hilbert-style system for the classical propositional logic CL (see, for instance, [43]).

This logic is known to be sound and complete with respect to the standard semantics of truth tables (see, e.g., [43]). This semantics is an integral part of the semantics for modal and justification logics considered in this book.

This semantics is also indispensable for the tableau proof system for CL. Tableaus are a natural proof system that, unlike many other proof systems, often serves as a decision algorithm. If a tableau proof for a formula A is impossible to obtain, it is often possible (for decidable logics) to continue a tableau construction long enough to determine that $\neg A$ can be satisfied. Moreover, a countermodel for A is usually easily read from the resulting tableau construction. This property of tableau systems makes them a popular tool used for decision procedures. Moreover, tableau-based algorithms often turn out to have the lowest possible complexity.

We are not planning to supply every logic we work with with a tableau system (although they do exist in most cases). However, the tableau proof system for CL, with a minor modification, will be used in Chapter 5. Hence, we now present it here.

Definition 1.11 (Signed formula). A signed formula has a form TA or FA where A is an ordinary formula.

Definition 1.12 (Propositional tableau). A *propositional tableau* is a finite binary tree that contains a signed formula in each node such that no branch contains the same signed formula twice.

Definition 1.13 (Open branch, open tableau). A branch of the tableau is called (propositionally) *closed* if it contains the signed formula $T\bot$ or if it contains formulas TA and FA for some A. A tableau is called *closed* if all its branches are closed. A branch (tableau) that is not closed is *open*.

Definition 1.14 (Tableau construction rules). *Tableau construction rules* are used to extend a given tableau. These rules are:

- If a formula $FA \to B$ is present on an open branch, this branch is appended with the formulas TA and FB (without creating duplicate formulas);

- If a formula $TA \to B$ is present on an open branch, this branch is split into two branches, one appended with the formulas FA and the other appended with the formula TB (without creating duplicate formulas on either branch).

Remark 1.15. The restriction of at most one formula copy per branch causing the "without creating duplicate formulas" qualification is implemented as follows. If two formulas need to be added to a branch of which exactly one is already present on the branch, then only the second formula is added. For instance, if a branch with $FA \to B$ already contains TA, then only FB is added. Moreover, if a branch with $FA \to B$ already contains both TA and FB, then the tableau construction rule cannot be applied to $FA \to B$.

Similarly, if a branch with $TA \to B$ already contains both FA and TB, then the tableau construction rule cannot be applied to $TA \to B$. Moreover, in this case, even if only one of FA and TB is already present on the branch, the rule is not applied. Suppose FA is present. In principle, we could apply this rule and create an extension of the branch with TB. But the second branch without TB must still be there, and to close the tableau both have to be closed. It is reasonably clear that the branch with TB can be closed whenever the one without TB can. Thus, it makes no sense to check the former condition separately.

Definition 1.16 (Tableau proof). A *tableau proof* of a formula A is a closed tableau that is obtained by repeatedly applying *tableau construction rules* to a single-node tableau with FA in its only node.

Definition 1.17 (Satisfiable branch). A tableau branch is called *satisfiable* if there exists a truth assignment that makes all T-signed formulas on the branch true and all F-signed formulas on the branch false. A tableau is *satisfiable* if one of its branches is.

1.1 Propositional Logic

It is easy to see that

Lemma 1.18. *A tableau construction rule applied to a satisfiable tableau yields a satisfiable tableau.*

Corollary 1.19 (Tableau completeness). *If A has a tableau proof it is a classical propositional tautology.*

Proof. Suppose towards a contradiction that A were not a tautology, i.e., the tableau consisting of FA were satisfiable. Then the closed tableau that proves A would also be satisfiable. However, all its branches are closed, and no closed branch can be satisfiable. □

Definition 1.20 (Completed tableau). A tableau is called *completed* if no tableau construction rule can be applied to it.

We will make use of König's Lemma [85, 125] to show that a completed tableau is constructed in finitely many steps.

Lemma 1.21 (König's Lemma). *Every infinite finitely branching tree has an infinite path.*

Lemma 1.22. *Given any initial tableau, any sequence of tableau construction rules applied to it leads to a completed tableau in finitely many steps.*

Proof. Whenever a signed formula TA or FA is added to a branch by a tableau construction rule the formula A is a subformula of another signed formula already present on the branch. Hence, the set of signed formulas that can occur on a given branch is finite: it contains at most twice the number of subformulas of the set of signed formulas originally present on the graph. Since all the branches of the tableau tree are finite, by König's Lemma, the tableau must be finite, i.e., it must be completed after finitely many steps. Moreover, the length of each branch in each completed tableau constructed from the initial one is linear in the size of the initial tableau because the set of subformulas of a given formula is linear in its size. □

Corollary 1.23 (Tableau soundness). *If A is a classical tautology, it has a tableau proof.*

Proof. Suppose a weaker property than not having a tableau proof: suppose that some tableau construction starting from FA does not lead to a tableau proof. We know that each tableau construction terminates, so it must then yield a completed open tableau, i.e., a completed tableau with an open branch. Define a truth assignment as follows: an atomic proposition P is made true iff TP is present on this open branch.

We show the two statements: that for each signed formula TB on the branch B is true and that for each signed formula FB on the branch B is false by simultaneous induction on the size of B. For TP on the branch this property holds by definition; for FP on the branch it holds because the branch is open and, hence, TP cannot be on it. If $F\bot$ is on the branch, we have $\bot$ false. $T\bot$ cannot be on an open branch. If $TB \to C$ is on the branch, then either FB or TC is because the tableau is completed and no rule can be applied. Thus, by the induction hypothesis either B is false or C is true. In either case, $B \to C$ is true. The case of $FB \to C$ on the branch is analogous.

Since the tableau started with FA, this signed formula is present on each branch, in particular, A is false for the constructed truth assignment and, hence, is not a tautology. □

1.1.2 Intuitionistic Propositional Logic

The language $\mathcal{L}_{\text{cp}}$ was chosen with a minimal complete set of Boolean connectives $\{\bot, \to\}$. *Completeness* here means that all possible Boolean functions, i.e., all other Boolean connectives, can be defined through the chosen connectives. Such definitions are given in the form of classically valid equivalences. Virtually all these equivalences cease to be valid under intuitionistic reasoning, which makes it necessary to expand the language.

Definition 1.24 (Intuitionistic propositional language). We assume the same set $\text{Prop} = \{P_0, P_1, \ldots, P_n, \ldots\}$ of *atomic propositions* as in $\mathcal{L}_{\text{cp}}$. Formulas of the language $\mathcal{L}_{\text{ip}}$ of intuitionistic propositional logic are defined inductively as follows:

1. $\bot$ is a formula of $\mathcal{L}_{\text{ip}}$;

2. each atomic proposition $P_i \in \text{Prop}$ is a formula of $\mathcal{L}_{\text{ip}}$;

3. if A and B are formulas of $\mathcal{L}_{\text{ip}}$, then so are

    - $(A \vee B)$, read A *or* B,

1.1 Propositional Logic

- $(A \wedge B)$, read *A and B*, and
- $(A \to B)$.

So conjunction and disjunction are primary connectives of $\mathcal{L}_{\mathrm{ip}}$, while *verum* $\top$, negation, and equivalence retain the definitions used in the classical case. As a result, the two languages, $\mathcal{L}_{\mathrm{cp}}$ and $\mathcal{L}_{\mathrm{ip}}$ look identical when defined connectives are allowed. This causes no confusion and is, in fact, helpful for explaining the relationship between the two logics. For instance, the statement that each intuitionistic tautology is also valid classically (see Lemma 1.48 below) involves an implicit translation from the intuitionistic language to the classical one. It is also possible to use the same language ($\mathcal{L}_{\mathrm{ip}}$) for both logics, but it would unnecessarily inflate the many proofs by induction on the formula construction that we employ in the book.

All the notational conventions, including the omission of unnecessary parentheses, described for $\mathcal{L}_{\mathrm{cp}}$ remain in effect for $\mathcal{L}_{\mathrm{ip}}$.

The *intuitionistic propositional logic* IPC is described by a Hilbert-style system with axiom schemes **PAx1**–**PAx9** and the inference rule *modus ponens* (**MP**) from Figure 1.2 (see, for instance, [41, Sect. 2.6]). The schemes **PAx1** and **PAx2** from CL are also schemes of IPC (however, now they consist of instances in the language $\mathcal{L}_{\mathrm{ip}}$). We need further schemes **PAx3**–**PAx9** to axiomatize the connectives for conjunction and disjunction. Finally we need **PAx9** to axiomatize the intuitionistic content of $\bot$ (note that **LDN** is not valid intuitionstically). We write IPC $\vdash A$ if a formula A is derivable in this Hilbert-style system.

We will describe the relationship between classical and intuitionistic propositional logics in Section 1.3. For more details we refer to [138].

1.1.3 Connectives of Variable Arity

Given the commutativity and associativity of the binary operations of conjunction and disjunction, it is standard to extend them to finite sets of formulas. Indeed, given a finite set $\Gamma = \{A_1, \ldots, A_n\}$ and a permutation $\pi \colon \{1, \ldots, n\} \to \{1, \ldots, n\}$ all well-formed formulas obtained by parenthesizing the expression

$$A_{\pi(1)} \wedge \cdots \wedge A_{\pi(n)}$$

are both classically and intuitionistically equivalent; the situation with disjunction is analogous. Thus, semantically we can talk about the conjunction and disjunction of the set Γ, to be denoted by $\bigwedge \Gamma$ and $\bigvee \Gamma$

PAx1 $P \to (Q \to P)$

PAx2 $(P \to (Q_1 \to Q_2)) \to ((P \to Q_1) \to (P \to Q_2))$

PAx3 $P \wedge Q \to P$

PAx4 $P \wedge Q \to Q$

PAx5 $P \to (Q \to P \wedge Q)$

PAx6 $P \to P \vee Q$

PAx7 $Q \to P \vee Q$

PAx8 $(P_1 \to Q) \to ((P_2 \to Q) \to (P_1 \vee P_2 \to Q))$

PAx9 $\bot \to P$

Inference rule *modus ponens* $\dfrac{A \quad A \to B}{B}$ (**MP**).

Figure 1.2: The axiom schemes and inference rule of the Hilbert-style system for the intuitionistic propositional logic IPC (see, for instance, [41, Sect. 2.6]).

respectively. However, syntactically, we should still define particular syntactic objects to represent $\bigwedge \Gamma$ and $\bigvee \Gamma$. For instance, in order to apply (**MP**) to formulas $\bigwedge \Gamma \to B$ and $\bigwedge \Gamma$, it is necessary that both occurrences of $\bigwedge \Gamma$ represent the same syntactic object, but it doesn't matter which particular conjunction is chosen. This enables us to employ the "arbitrary but fixed" trick.

We assume that all symbols of all the alphabets used in this book for writing formulas are ordered (without specifying the ordering). This ordering $\prec$ is fixed throughout the book. This ordering induces a lexicographic $\prec$ order on all words in the union of the alphabets used in the book, including on the well-formed formulas of all the languages we use. Given a set $\Gamma = \{A_1, \ldots, A_n\}$ of formulas in one of these languages, let $\pi \colon \{1, \ldots, n\} \to \{1, \ldots, n\}$ be the permutation that orders the formulas from Γ according to $\prec$, i.e., such that

$$A_{\pi(1)} \preceq A_{\pi(2)} \preceq \cdots \preceq A_{\pi(n)} \;. \tag{1.1}$$

We define
$$\bigwedge \Gamma := \Big(\big((A_{\pi(1)} \wedge A_{\pi(2)}) \wedge A_{\pi(3)}\big) \dots \Big) \wedge A_{\pi(n)} \, ,$$
$$\bigvee \Gamma := \Big(\big((A_{\pi(1)} \vee A_{\pi(2)}) \vee A_{\pi(3)}\big) \dots \Big) \vee A_{\pi(n)} \, .$$

It is straightforward to extend this definition to the case when Γ is a multiset rather than a set. The only difference is that there may be more than one permutation satisfying 1.1, but all such permutations produce the same formula. For $\Gamma = \varnothing$, we define
$$\bigwedge \varnothing := \top \quad \text{and} \quad \bigvee \varnothing := \bot \, .$$

All of our languages share certain propositional symbols. Thus, some syntactic expressions are well-formed formulas in several languages at the same time. In particular, Γ may be a (multi)set of formulas in each of several languages. By using a global lexicographic order, we ensure that $\bigwedge \Gamma$ and $\bigvee \Gamma$ represent the same formula no matter within the context of which language Γ is considered. This makes $\bigwedge \Gamma$ and $\bigvee \Gamma$ well-defined global Boolean operations on finite sets of formulas.

Remark 1.25. These operations should be applied with caution to formulas viewed both classically and intuitionistically. For instance, if we consider $\Gamma \subseteq \mathcal{L}_{\mathrm{ip}}$ with the formula $C = A \vee B$ as a set of classical formulas, then C becomes $A \to \bot \to B$ and its place in the lexicographic order may change, causing a possible permutation in $\bigwedge \Gamma$ and $\bigvee \Gamma$.

1.2 Modal Logic

Modal logic is an extension of propositional logic with so-called *modalities*. As the name suggests, a modality modifies the meaning of the statement it is applied to. If a formula A is to represent the truth of a certain proposition, then, for a modality $\heartsuit$, the formula $\heartsuit A$ could represent any of the following attitudes, depending on which flavor of modal logic is employed:

- A is necessarily true,
- A is obligatory,
- A will always be true,

- A is believed to be true.

There are many more attitudes that can be viewed as modalities. It is also possible to consider several such attitudes within one language. In particular, it is quite common to always consider the dual modality $\neg\heartsuit\neg$. For instance, if $\heartsuit A$ stands for A is necessarily true, then $\neg\heartsuit\neg A$ means that it is possible that A is true. Similarly, if $\heartsuit A$ stands for A is known, then $\neg\heartsuit\neg A$ means that it is possible, for all that is known, that A, see [78].

There are traditional names given to some types of modalities:

- necessity is almost always $\Box$,

- possibility is almost always $\Diamond$,

- belief is often B, with its dual denoted as $\hat{B}$,

- knowledge is often K, with its dual denoted as $\hat{K}$,

- it is also common to write either of belief or knowledge as $\Box$.

However, the knowledge of the meaning assigned to modalities is only necessary to interpret formulas into natural language. Reasoning about them formally can be done in a manner independent of the particular shade of modality. This will be the most common mode of operation in this book. We usually work with two modalities: $\Box$ and its dual $\Diamond = \neg\Box\neg$.

Modalities require additional postulates describing their properties. Naturally, these postulates depend on the kind of attitude one intends to represent and often also on his/her philosophical inclinations. We rarely go into the motivations justifying the choice of modal postulates. In fact, for the most part we concentrate on one particular set of modal postulates, called S4, to be soon introduced.

In principle, any treatment of modalities can be combined with either the classical or intuitionistic treatment of propositional logic. In this book, we only consider classical modal logics.

1.2.1 Modal Logic: Syntax

Definition 1.26 (Modal language). We assume the same set

$$\mathsf{Prop} = \{P_0, P_1, \ldots, P_n, \ldots\}$$

1.2 Modal Logic

of *atomic propositions* as in $\mathcal{L}_{\mathrm{cp}}$ and $\mathcal{L}_{\mathrm{ip}}$. *Formulas of the language $\mathcal{L}_\Box$ of modal logic* are defined inductively as follows:

1. $\bot$ is a formula of $\mathcal{L}_\Box$;
2. each atomic proposition $P_i \in \mathsf{Prop}$ is a formula of $\mathcal{L}_\Box$;
3. if A and B are formulas of $\mathcal{L}_\Box$, then so are
   - $(A \to B)$ and
   - $\Box A$, read *box A*.

Remark 1.27. The modal language $\mathcal{L}_\Box$ is an extension of the classical propositional language $\mathcal{L}_{\mathrm{cp}}$, i.e., $\mathcal{L}_{\mathrm{cp}} \subseteq \mathcal{L}_\Box$.

To the Boolean connectives $\neg, \top, \lor, \land,$ and $\leftrightarrow$ defined as in $\mathcal{L}_{\mathrm{cp}}$, we add a new abbreviation:

- $\Diamond A := \neg\Box\neg A$, read *diamond A*.

Similar to negation $\neg$ the two unary modal operators $\Box$ and $\Diamond$ bind stronger than all binary connectives. The agreements about omitting parentheses remain unchanged. Set conjunction and disjunction are defined as before for the propositional languages. In addition, we define two operations on sets of modal formulas. For a set $\Gamma \subseteq \mathcal{L}_\Box$, we define

$$\Box\Gamma := \{\Box A \mid A \in \Gamma\},$$
$$\Diamond\Gamma := \{\Diamond A \mid A \in \Gamma\}.$$

These operations can also be applied to multisets.

The subformulas of an $\mathcal{L}_\Box$-formula are defined as usual.

Definition 1.28 (Subformula). The set of subformulas $\mathsf{sub}(F)$ of an $\mathcal{L}_\Box$-formula F is inductively defined by:

1. $\mathsf{sub}(\bot) := \{\bot\}$,
2. $\mathsf{sub}(P_i) := \{P_i\}$ for each $P_i \in \mathsf{Prop}$,
3. $\mathsf{sub}(A \to B) := \{A \to B\} \cup \mathsf{sub}(A) \cup \mathsf{sub}(B)$,
4. $\mathsf{sub}(\Box A) := \{\Box A\} \cup \mathsf{sub}(A)$.

The *modal logic* S4 is described by a Hilbert-style system with all axiom schemes of CL (see Figure 1.1) and modal axiom schemes from Figure 1.3 and with the inference rules (**MP**) (see Figure 1.1) and *necessitation* (**Nec**) from Figure 1.3. We write $\mathsf{S4} \vdash A$ if a formula A is derivable in this Hilbert-style system.

k $\Box(P \to Q) \to (\Box P \to \Box Q)$

t $\Box P \to P$

4 $\Box P \to \Box\Box P$

Inference rule *necessitation* $\dfrac{A}{\Box A}$ (**Nec**).

Figure 1.3: The modal axiom schemes and inference rule of the Hilbert-style system for the modal logic S4.

1.2.2 Propositional Reasoning

Every formula representing an axiom scheme of CL also represents an axiom scheme of S4. Since, in addition, $\mathcal{L}_{\text{cp}} \subseteq \mathcal{L}_\Box$, it follows that every axiom of CL is also an axiom of S4. Modus ponens, the only inference rule of CL, is also an inference rule of S4. We will describe such situations by simply saying that the axiom system S4 is an *extension* of the axiom system CL. It is an easy corollary that CL $\subseteq$ S4. Moreover, we have the following theorem:

Theorem 1.29 (Conservativity of S4 over CL). *The logic S4 is a conservative extension of CL, i.e., for any formula* $A \in \mathcal{L}_{cp}$,

$$\text{S4} \vdash A \quad \Longleftrightarrow \quad \text{CL} \vdash A \ .$$

Proof. Since S4 is an extension of CL, we only need to show conservativity, i.e., the left-to-right implication. Consider the translation $(\cdot)^t : \mathcal{L}_\Box \to \mathcal{L}_{\text{cp}}$ defined by induction on the structure of the modal formula:

1. $P^t := P$;
2. $\bot^t := \bot$;
3. $(B \to C)^t := B^t \to C^t$;
4. $(\Box B)^t := B^t$.

This translation erases all modalities in a given formula. We prove by induction on the length of the derivation of A in S4 that CL $\vdash A^t$. We remind that a derivation of A in S4 is a finite list of $\mathcal{L}_\Box$-formulas with A at the end of the list.

So our induction hypothesis is that CL $\vdash B^t$ holds for any formula B that has an S4-derivation of length strictly less than n. Let A have a

1.2 Modal Logic

derivation of length n. There are four possibilities regarding what the last step of the derivation, the step yielding A, is. We consider each of the possibilities in turn:

1. If A is an $\mathcal{L}_\Box$-instance of a propositional axiom scheme from Figure 1.1, then A^t is an $\mathcal{L}_{\text{cp}}$-instance of the same axiom scheme.

2. If A is an instance of the **k**, **t**, or **4** axiom scheme, then A^t is a classical propositional tautology of the form $C \to C$.

3. If A is obtained by (**MP**) from B and $B \to A$ occurring in the given S4-derivation before A, then both B and $B \to A$ have S4-derivations of length less than n and, hence, by the induction hypothesis, $\mathsf{CL} \vdash B^t$ and $\mathsf{CL} \vdash B^t \to A^t$. Now A^t can be derived in CL from B^t and $B^t \to A^t$ by modus ponens.

4. If $A = \Box B$ is obtained by (**Nec**) from B occurring in the given S4-derivation before A, then $A^t = B^t$ and B has an S4-derivation of length less than n meaning that $\mathsf{CL} \vdash B^t$ by the induction hypothesis.

It remains to note that $A^t = A$ for each formula $A \in \mathcal{L}_{\text{cp}}$. $\qquad\square$

Remark 1.30. In future proofs we will omit the mention of the actual induction parameter n. Whenever a formula is obtained by an inference rule in any Hilbert-style proof system, the premise(s) of this rule must have derivations of smaller length, which enables us to apply the induction hypothesis to it (them). Thus, it is sufficient to prove the induction statement for all axioms and conclusions of zero-premise rules and show that the induction statement holds for the conclusion of any other rule whenever it holds for the premise(s).

So each classical propositional tautology is a theorem of S4. Moreover, each substitution instance of a classical propositional tautology is a theorem of S4: if $\mathsf{CL} \vdash A$, then $\mathsf{S4} \vdash A\sigma$ for any substitution σ of $\mathcal{L}_\Box$-formulas for atomic propositions. Indeed, to derive $A\sigma$ in S4, it is sufficient to apply the substitution σ to every formula in a derivation of A in CL. In this case, we can say that $A\sigma$ can be derived *using propositional reasoning* because neither instances of the modal axioms from Figure 1.3 nor the necessitation rule (**Nec**) are used in such a derivation.

Definition 1.31. For a logic L, a formula A *is derivable from a set* Σ *of formulas by propositional reasoning* if there exists a derivation of A

from Σ in L that uses instances of only those axiom schemes that are present in Figure 1.1 and only one inference rule, (**MP**).

In this book, we typically omit propositional reasoning trusting the reader to be able to fill in the details. Here is an example:

Lemma 1.32. *For any finite set* $\Sigma \subseteq \mathcal{L}_\Box$,

$$\mathsf{S4} \vdash \bigwedge \Sigma \to A \quad \text{implies} \quad \mathsf{S4} \vdash \bigwedge \Box\Sigma \to \Box A \ . \tag{1.2}$$

Proof. We show (1.2) by induction on the cardinality n of the set Σ. If $n = 0$, the premise is $\mathsf{S4} \vdash \top \to A$. By propositional reasoning,

$$\mathsf{S4} \vdash A$$

(indeed, $\top$ is an abbreviation of the formula $\bot \to \bot$, which is a classical propositional tautology and, hence, a theorem of $\mathsf{S4}$; we obtain A by (**MP**) from this theorem and $\top \to A$, which we assume to be derivable). By the rule (**Nec**),

$$\mathsf{S4} \vdash \Box A \ .$$

By propositional reasoning,

$$\mathsf{S4} \vdash \top \to \Box A \tag{1.3}$$

(indeed, $\Box A \to (\top \to \Box A)$ is an axiom, an instance of **PAx1**; we obtain $\top \to \Box A$ by (**MP**) from this axiom and $\Box A$). It remains to note that (1.3) is the conclusion of the lemma since $\bigwedge \Box \varnothing = \bigwedge \varnothing = \top$.

In the rest of the book, we omit the detailed descriptions of propositional reasoning such as presented in parentheses above.

For $n > 0$, let $B \in \Sigma$ and $\Sigma' := \Sigma \setminus \{B\}$. From the premise, by propositional reasoning, we get

$$\mathsf{S4} \vdash \bigwedge \Sigma' \to (B \to A) \ .$$

By the induction hypothesis, we obtain

$$\mathsf{S4} \vdash \bigwedge \Box\Sigma' \to \Box(B \to A) \ .$$

By the axiom scheme k and propositional reasoning, we find

$$\mathsf{S4} \vdash \bigwedge \Box\Sigma' \to (\Box B \to \Box A) \ .$$

Again by propositional reasoning, this yields

$$\mathsf{S4} \vdash \bigwedge \Box\Sigma \to \Box A$$

because $\Box\Sigma = \Box\Sigma' \cup \{\Box B\}$. $\square$

1.2 Modal Logic

In the second half of this proof, we use term "propositional reasoning" not only to omit tedious details, but also to avoid equally tedious case analysis. Indeed, the transition from $\bigwedge \Sigma \to A$ to $\bigwedge \Sigma' \to (B \to A)$ depends on the exact position of the conjunct B within the conjunction $\bigwedge \Sigma$. Thus, to spell out the propositional reasoning alluded to in the proof would require either considering multiple cases or insisting that B be, say, the lexicographically last member of Σ.

1.2.3 Modal Logic: Semantics

We introduce the *possible world semantics* (also called *Kripke semantics*) for modal logics, where a formula $\Box A$ is true at a world w, if and only if A is true at each world possible from the point of view of w. The relation "a world v is possible from the point of view of a world w" is modeled by a binary *accessibility* relation.

Definition 1.33. A binary relation $R \subseteq W \times W$ on the set W is called

- *reflexive* if $(u, u) \in R$ for each $u \in W$;
- *transitive* if $(u, v) \in R$ and $(v, w) \in R$ imply $(u, w) \in R$ for arbitrary $u, v, w \in W$.

Notation 1.34. In the context of Kripke semantics, it is standard to write $(w, v) \in R$, or $R(w, v)$, or wRv interchangeably. All three statements describe the situation when v is possible from the point of view of w.

Definition 1.35 (Kripke model). A *Kripke model* $\mathcal{M} = (W, R, \mathsf{val})$ for S4, or simply a Kripke model, is a triple where

1. W is a non-empty set,

2. $R \subseteq W \times W$ is a binary relation on W that is transitive and reflexive,

3. val is a valuation function $\mathsf{val}\colon \mathsf{Prop} \to \mathcal{P}(W)$, where $\mathcal{P}(X)$ denotes the power set of a set X.

The binary relation R is called the accessibility relation.

Remark 1.36. The requirement that R be transitive and reflexive is not part of the definition of Kripke models in general. We make it part of our definition because we work exclusively with S4.

We define what it means for a formula to hold at a world in a Kripke model inductively on the structure of the formula.

Definition 1.37 (Truth in a Kripke model)**.** Let $\mathcal{M} = (W, R, \mathsf{val})$ be a Kripke model and A be a formula of $\mathcal{L}_\Box$. For $w \in W$, we define $\mathcal{M}, w \Vdash A$ inductively by

1. for $A = \bot$, it is not the case that $\mathcal{M}, w \Vdash \bot$,

2. for $A = P \in \mathsf{Prop}$, we have $\mathcal{M}, w \Vdash P$ iff $w \in \mathsf{val}(P)$,

3. for $A = B \to C$, we have $\mathcal{M}, w \Vdash B \to C$ iff it is not the case that $\mathcal{M}, w \Vdash B$ or it is the case that $\mathcal{M}, w \Vdash C$,

4. for $A = \Box B$, we have $\mathcal{M}, w \Vdash \Box B$ iff $\mathcal{M}, v \Vdash B$ for all $v \in W$ with $R(w, v)$.

When $\mathcal{M}, w \Vdash A$, we say that the formula A holds at the world w of the model $\mathcal{M}$. We write $\mathcal{M}, w \nVdash A$ and say that the formula A does not hold at the world w of model $\mathcal{M}$ if it is not the case that $\mathcal{M}, w \Vdash A$. We write $\mathcal{M} \Vdash A$ and say that the formula A is *valid in the model* $\mathcal{M}$ if $\mathcal{M}, w \Vdash A$ for all $w \in W$. A formula A is called S4-*valid*, or simply valid, if for all Kripke models $\mathcal{M}$ for S4 we have $\mathcal{M} \Vdash A$.

Corollary 1.38. *It immediately follows from the definition of truth in Kripke models that*

1. $\mathcal{M}, w \Vdash \neg B$ *iff* $\mathcal{M}, w \nVdash B$;

2. $\mathcal{M}, w \Vdash \top$ *for any model* $\mathcal{M} = (W, R, \mathsf{val})$ *and any world* $w \in W$;

3. $\mathcal{M}, w \Vdash B \lor C$ *iff* $\mathcal{M}, w \Vdash B$ *or* $\mathcal{M}, w \Vdash C$;

4. $\mathcal{M}, w \Vdash B \land C$ *iff* $\mathcal{M}, w \Vdash B$ *and* $\mathcal{M}, w \Vdash C$;

5. $\mathcal{M}, w \Vdash B \leftrightarrow C$ *iff either*

    - $\mathcal{M}, w \Vdash B$ *and* $\mathcal{M}, w \Vdash C$ *or*
    - $\mathcal{M}, w \nVdash B$ *and* $\mathcal{M}, w \nVdash C$;

6. *for a finite set* $\Gamma \subseteq \mathcal{L}_\Box$*, we have* $\mathcal{M}, w \Vdash \bigwedge \Gamma$ *iff* $\mathcal{M}, w \Vdash B$ *for each formula* $B \in \Gamma$;

7. *for a finite set* $\Gamma \subseteq \mathcal{L}_\Box$*, we have* $\mathcal{M}, w \Vdash \bigvee \Gamma$ *iff* $\mathcal{M}, w \Vdash B$ *for some formula* $B \in \Gamma$;

1.2 Modal Logic

8. $\mathcal{M}, w \Vdash \Diamond B$ iff $\mathcal{M}, v \Vdash B$ for some $v \in W$ with $R(w,v)$.

Theorem 1.39 (Soundness). *Let A be a formula of $\mathcal{L}_\Box$. We have that*

$$\mathsf{S4} \vdash A \quad \text{implies} \quad A \text{ is valid.}$$

Proof. By induction on the length of the derivation of A we show that A is valid, i.e., for all Kripke models $\mathcal{M} = (W, R, \mathsf{val})$ and all $w \in W$ we have $\mathcal{M}, w \Vdash A$. We only show the validity of modal axioms from Figure 1.3 and that the necessitation rule preserves vaildity. The other cases work the same way as for CL.

1. Instances of the axiom scheme **k**: $A = \Box(B \to C) \to (\Box B \to \Box C)$. Assume (a) $\mathcal{M}, w \Vdash \Box(B \to C)$ and (b) $\mathcal{M}, w \Vdash \Box B$. We have to show $\mathcal{M}, w \Vdash \Box C$, that is (c) $\mathcal{M}, v \Vdash C$ for all $v \in W$ with $R(w,v)$. From (b) we get $\mathcal{M}, v \Vdash B$ and from (a) we get $\mathcal{M}, v \Vdash B \to C$ for all $v \in W$ with $R(w,v)$. Thus we conclude (c).

2. Instances of the axiom scheme **t**: $A = \Box B \to B$. Assume $\mathcal{M}, w \Vdash \Box B$, that is $\mathcal{M}, v \Vdash B$ for all $v \in W$ with $R(w,v)$. Since R is reflexive, we have $R(w,w)$, and we conclude $\mathcal{M}, w \Vdash B$.

3. Instances of the axiom scheme **4**: $A = \Box B \to \Box\Box B$. Assume

$$\mathcal{M}, w \Vdash \Box B \ . \tag{d}$$

We have to show $\mathcal{M}, w \Vdash \Box\Box B$, that means $\mathcal{M}, u \Vdash B$ for arbitrary $u, v \in W$ with $R(w,v)$ and $R(v,u)$. We have $R(w,u)$ since R is transitive. Hence, $\mathcal{M}, u \Vdash B$ by (d).

4. $A = \Box B$ is inferred from B by the necessitation rule (**Nec**). By the induction hypothesis, B is valid. In particular, it means $\mathcal{M}, v \Vdash B$ for all $v \in W$ with $R(w,v)$. Thus we conclude $\mathcal{M}, w \Vdash \Box B$. □

To show completeness of S4 we make use of a so-called canonical model, which is based on maximal consistent sets. This construction is very general and will also be employed later to show completeness for justification logics.

The following definitions are not given for S4 only, but more general, for any logic L with classical propositional logic in the background, i.e., with all instances of axiom schemes from Figure 1.1 being derivable and with the inference rule (**MP**) being admissible.

Definition 1.40 (Consistency). Let L be a logic in a language $\mathcal{L}$. A set $\Gamma \subseteq \mathcal{L}$ of formulas is L-*consistent* if
$$\mathsf{L} \nvdash \bigwedge \Sigma \to \bot$$
for each finite subset $\Sigma \subseteq \Gamma$.

A set Γ is called *maximal* L-*consistent* if it is consistent and none of its proper supersets is.

It is trivial to see that any subset of an L-consistent set is itself L-consistent. Equivalently, if a set is not L-consistent, none of its supersets is.

Lemma 1.41 (Lindenbaum Lemma). *For each L-consistent set Δ there exists a maximal L-consistent set Γ such that $\Delta \subseteq \Gamma$.*

Proof. Let Δ be an L-consistent set. Take any fixed enumeration
$$A_0, A_1, \ldots, A_n, \ldots$$
of all formulas of the language $\mathcal{L}$. We set

$$\Delta_0 := \Delta \ ;$$
$$\Delta_{n+1} := \begin{cases} \Delta_n \cup \{A_n\} & \text{if } \Delta_n \cup \{A_n\} \text{ is L-consistent,} \\ \Delta_n & \text{otherwise;} \end{cases}$$
$$\Gamma := \bigcup_{n \in \omega} \Delta_n \ .$$

It follows from the construction of the sets Δ_n that
$$\Delta_0 \subseteq \Delta_1 \subseteq \cdots \subseteq \Delta_n \subseteq \ldots$$
is an increasing sequence of L-consistent sets. Their union Γ is also L-consistent. Consider an arbitrary finite subset $\Sigma = \{A_{i_1}, \ldots, A_{i_k}\}$ of Γ. Each formula $A_{i_j} \in \Sigma$ is taken from some set Δ_{n_j}. Let
$$N := \max\{n_1, \ldots, n_k\} \ .$$
Then $\Delta_{n_j} \subseteq \Delta_N$ for each $j = 1, \ldots, k$. Thus, $\Sigma \subseteq \Delta_N$ and $\mathsf{L} \nvdash \bigwedge \Sigma \to \bot$ by the L-consistency of Δ_N.

Moreover, Γ is maximal L-consistent. Indeed, for any $A_j \notin \Gamma$, clearly $A_j \notin \Delta_{j+1}$ since $\Delta_{j+1} \subseteq \Gamma$. This means that $\Delta_j \cup \{A_j\}$ is not L-consistent, and, consequently, neither is its superset $\Gamma \cup \{A_j\}$. Since it is not possible to extend Γ by any single formula in an L-consistent manner, no proper superset of Γ is L-consistent.

It remains to note that $\Delta = \Delta_0 \subseteq \Gamma$ by construction. □

1.2 Modal Logic

Lemma 1.42. *Let Γ be a maximal L-consistent set.*

1. *$A \in \Gamma$ for all formulas A such that $\mathsf{L} \vdash A$.*

2. *$A \in \Gamma$ if and only if $\neg A \notin \Gamma$.*

3. *$A \to B \in \Gamma$ if and only if $A \notin \Gamma$ or $B \in \Gamma$.*

4. *$A \in \Gamma$ and $A \to B \in \Gamma$ imply $B \in \Gamma$, i.e., maximal L-consistent sets are closed with respect to* (**MP**).

Proof. Since we only consider logics L with classical propositional logic in the background, we can freely use propositional reasoning in the proof.

1. Let $\mathsf{L} \vdash A$. Suppose towards a contradiction that $A \notin \Gamma$. By the maximal consistency of Γ this would imply that $\Gamma \cup \{A\}$ is not L-consistent, i.e., that there exists a finite set $\Sigma' \subseteq \Gamma \cup \{A\}$ such that
$$\mathsf{L} \vdash \bigwedge \Sigma' \to \bot \ .$$

   By propositional reasoning,
$$\mathsf{L} \vdash A \wedge \bigwedge \Sigma' \to \bot \ .$$

   For a set $\Sigma := \Sigma' \setminus \{A\} \subseteq \Gamma$, by propositional reasoning,
$$\mathsf{L} \vdash A \wedge \bigwedge \Sigma \to \bot \ . \tag{1.4}$$

   Since $\mathsf{L} \vdash A$, propositional reasoning would yield
$$\mathsf{L} \vdash \bigwedge \Sigma \to \bot \ ,$$

   which would mean that Γ is not L-consistent and would contradict the assumption that Γ is maximal L-consistent.

2. First we show that, for each formula $A \in \mathcal{L}$, at least one of the formulas A and $\neg A$ belongs to Γ. Suppose towards a contradiction that $A \notin \Gamma$ and $\neg A \notin \Gamma$. Then neither $\Gamma \cup \{A\}$ nor $\Gamma \cup \{\neg A\}$ would be L-consistent by the maximal consistency of Γ. Using the argument employed earlier to conclude (1.4) in part 1 of this lemma, we conclude that there would exist finite sets $\Sigma_1, \Sigma_2 \subseteq \Gamma$ such that
$$\mathsf{L} \vdash A \wedge \bigwedge \Sigma_1 \to \bot$$

and
$$L \vdash \neg A \wedge \bigwedge \Sigma_2 \to \bot \ .$$
Thus, by propositional reasoning, for a set $\Sigma := \Sigma_1 \cup \Sigma_2 \subseteq \Gamma$,
$$L \vdash (A \vee \neg A) \wedge \bigwedge \Sigma \to \bot \ .$$
Further, since $L \vdash A \vee \neg A$ by propositional reasoning, we would have
$$L \vdash \bigwedge \Sigma \to \bot \ ,$$
which would mean that Γ is not L-consistent and would contradict the assumption that Γ is maximal L-consistent.

It remains to show that A and $\neg A$ cannot both belong to Γ. Suppose towards a contradiction that $\{A, \neg A\} \subseteq \Gamma$. Since, by propositional reasoning,
$$L \vdash A \wedge \neg A \to \bot \ ,$$
this would mean that Γ is not L-consistent in contradiction to the assumptions.

3. Let
$$A \notin \Gamma \quad \text{or} \quad B \in \Gamma \ . \tag{1.5}$$
Suppose towards a contradiction that $A \to B \notin \Gamma$. Since Γ is maximal consistent, there would exist a finite set $\Sigma \subseteq \Gamma$ such that
$$L \vdash (A \to B) \wedge \bigwedge \Sigma \to \bot \ .$$
By propositional reasoning, this would imply
$$L \vdash \neg A \wedge \bigwedge \Sigma \to \bot$$
and
$$L \vdash B \wedge \bigwedge \Sigma \to \bot \ .$$
Thus, neither $\Gamma \cup \{\neg A\}$ nor $\Gamma \cup \{B\}$ could be L-consistent. Since Γ is L-consistent by the assumptions, neither $\neg A$ nor B could belong to Γ. From this and (1.5), it would follow that $A \notin \Gamma$ and $\neg A \notin \Gamma$, thus violating part 2 of this lemma.

Let $A \to B \in \Gamma$. Suppose towards a contradiction that $A \in \Gamma$ and $B \notin \Gamma$. Then, $\neg A \notin \Gamma$ by part 2 of this lemma. By the maximal consistency of Γ, there would exist finite sets $\Sigma_1, \Sigma_2 \subseteq \Gamma$ such that
$$L \vdash \neg A \wedge \bigwedge \Sigma_1 \to \bot$$

1.2 Modal Logic

and
$$\mathsf{L} \vdash B \wedge \bigwedge \Sigma_2 \to \bot \ ,$$
which would imply by propositional reasoning, for
$$\Sigma := \Sigma_1 \cup \Sigma_2 \subseteq \Gamma \ ,$$
that
$$\mathsf{L} \vdash (A \to B) \wedge \bigwedge \Sigma \to \bot \ .$$
This would mean that $\Gamma \cup \{A \to B\}$ is not L-consistent. Because of $A \to B \in \Gamma$, it would follow that Γ is not L-consistent in contradiction to the assumption of maximal L-consistency of Γ.

4. This is a reformulation of the direction from left to right from part 3 of this lemma. □

Additional properties of maximal L-consistent sets can either be proved analogously or derived from the preceding lemma.

Corollary 1.43. *Let Γ be a maximal L-consistent set.*

1. $\bot \notin \Gamma$.

2. $A \wedge B \in \Gamma$ *if and only if* $A \in \Gamma$ *and* $B \in \Gamma$.

3. $A \vee B \in \Gamma$ *if and only if* $A \in \Gamma$ *or* $B \in \Gamma$.

We now return to the specific case of the modal language and the logic S4.

Definition 1.44 (Canonical model). For a set $\Gamma \subseteq \mathcal{L}_\Box$ of formulas, we define
$$\Gamma/\Box := \{A \mid \Box A \in \Gamma\} \ .$$
The *canonical model* $\mathcal{M}^c := (W^c, R^c, \mathsf{val}^c)$ for the logic S4, or simply the canonical model, is given by:

1. $W^c := \{\Gamma \subseteq \mathcal{L}_\Box \mid \Gamma \text{ is a maximal S4-consistent set}\}$;

2. $R^c := \{(\Gamma, \Delta) \in W^c \times W^c \mid \Gamma/\Box \subseteq \Delta\}$;

3. $\mathsf{val}^c \colon \mathsf{Prop} \to \mathcal{P}(W^c)$ is a function defined by
$$\mathsf{val}^c(P) := \{\Gamma \in W^c \mid P \in \Gamma\} \ . \tag{1.6}$$

Lemma 1.45. *The canonical model $\mathcal{M}^c = (W^c, R^c, \mathsf{val}^c)$ for $\mathsf{S4}$ is a Kripke model for $\mathsf{S4}$.*

Proof. There are three things to be demonstrated: that W^c is not empty, that R^c is reflexive, and that R^c is transitive.

1. To show that W^c is not empty, it is sufficient to demonstrate that the empty set $\varnothing$ is $\mathsf{S4}$-consistent. Then by the Lindenbaum lemma, there exists a maximal $\mathsf{S4}$-consistent superset of $\varnothing$, which is an element of W^c. Suppose towards a contradiction that $\varnothing$ is not $\mathsf{S4}$-consistent. This would mean that

   $$\mathsf{S4} \vdash \bigwedge \varnothing \to \bot \, ,$$

   in other words, $\mathsf{S4} \vdash \top \to \bot$. By conservativity of $\mathsf{S4}$ over CL, we would have $\mathsf{CL} \vdash \top \to \bot$, which would contradict the soundness of classical propositional logic with respect to truth tables.

2. To show that R^c is reflexive, we need to show that $\Gamma/\square \subseteq \Gamma$ for each $\Gamma \in W^c$. Let $A \in \Gamma/\square$, i.e., $\square A \in \Gamma$. Since Γ is maximal $\mathsf{S4}$-consistent and $\square A \to A$ is an axiom of $\mathsf{S4}$, it follows by Lemma 1.42.1 that $\square A \to A \in \Gamma$. We conclude $A \in \Gamma$ by Lemma 1.42.4.

3. To show that R^c is transitive, let $R^c(\Gamma, \Delta)$ and $R^c(\Delta, \Phi)$ for arbitrary $\Gamma, \Delta, \Phi \in W^c$. We need to show that $R^c(\Gamma, \Phi)$, i.e., that $\Gamma/\square \subseteq \Phi$. Assume $A \in \Gamma/\square$, in other words, $\square A \in \Gamma$. Since Γ is a maximal $\mathsf{S4}$-consistent set and $\square A \to \square\square A$ is an axiom of $\mathsf{S4}$, this axiom belongs to Γ by Lemma 1.42.1, and we infer $\square\square A \in \Gamma$ by Lemma 1.42.4. Thus, $\square A \in \Gamma/\square \subseteq \Delta$, from which we conclude that $A \in \Delta/\square \subseteq \Phi$. $\square$

Lemma 1.46 (Truth lemma). *Let $\mathcal{M}^c = (W^c, R^c, \mathsf{val}^c)$ be the canonical model. For each world $\Gamma \in W^c$ and each formula $A \in \mathcal{L}_\square$ we have*

$$\mathcal{M}^c, \Gamma \Vdash A \quad \text{if and only if} \quad A \in \Gamma \, .$$

Proof. By induction on the structure of A.

1. A is an atomic proposition P. We have $\mathcal{M}^c, \Gamma \Vdash P$ iff $\Gamma \in \mathsf{val}^c(P)$ iff $P \in \Gamma$.

2. $A = \bot$. We have $\mathcal{M}^c, \Gamma \nVdash \bot$ and $\bot \notin \Gamma$ for any $\Gamma \in W^c$.

1.2 Modal Logic

3. $A = B \to C$. We have $\mathcal{M}^c, \Gamma \Vdash B \to C$ iff $\mathcal{M}^c, \Gamma \nVdash B$ or $\mathcal{M}^c, \Gamma \Vdash C$ iff (by the induction hypothesis) $B \notin \Gamma$ or $C \in \Gamma$ iff (by Lemma 1.42.3) $B \to C \in \Gamma$.

4. $A = \Box B$. To show the right-to-left direction, assume $\Box B \in \Gamma$. Hence, $B \in \Gamma/\Box$. For any $\Delta \in W^c$ such that $R^c(\Gamma, \Delta)$, we have $B \in \Delta$ by the definition of R^c and, consequently, $\mathcal{M}^c, \Delta \Vdash B$ by the induction hypothesis. We conclude that $\mathcal{M}^c, \Gamma \Vdash \Box B$.

   To show the left-to-right direction, assume $\mathcal{M}^c, \Gamma \Vdash \Box B$. Then

$$\Gamma/\Box \cup \{\neg B\} \text{ is not S4-consistent.} \tag{1.7}$$

Suppose towards a contradiction that (1.7) does not hold. Then, by the Lindenbaum lemma, we would find a maximal S4-consistent set $\Delta \supseteq \Gamma/\Box \cup \{\neg B\}$ and, by the definition of R^c, we would have $R^c(\Gamma, \Delta)$. By Lemma 1.42.2, $B \notin \Delta$. Using the induction hypothesis, we would have $\mathcal{M}^c, \Delta \nVdash B$, which would contradict our assumption that $\mathcal{M}^c, \Gamma \Vdash \Box B$.

The now established (1.7) implies that there exists a finite set $\Sigma \subseteq \Gamma/\Box$ such that

$$\mathsf{S4} \vdash \neg B \wedge \bigwedge \Sigma \to \bot \ .$$

By propositional reasoning, this implies

$$\mathsf{S4} \vdash \bigwedge \Sigma \to B \ .$$

By Lemma 1.32, we get

$$\mathsf{S4} \vdash \bigwedge \Box \Sigma \to \Box B \ ,$$

which by Lemma 1.42.1 implies

$$\bigwedge \Box \Sigma \to \Box B \in \Gamma \ .$$

Because $\Box \Sigma \subseteq \Gamma$, by Corollary 1.43.2 and induction on the cardinality of Σ we get

$$\bigwedge \Box \Sigma \in \Gamma \ .$$

Hence by Lemma 1.42.4 we conclude $\Box B \in \Gamma$ as required. □

Theorem 1.47 (Completeness). *Let A be a formula of $\mathcal{L}_\square$.*

$$A \text{ is valid} \quad \text{implies} \quad \mathsf{S4} \vdash A \ .$$

Proof. We show the contraposition. Assume $\mathsf{S4} \nvdash A$. By propositional reasoning, $\mathsf{S4} \nvdash \neg A \to \bot$. It follows that $\{\neg A\}$ is $\mathsf{S4}$-consistent and, hence, contained in some maximal $\mathsf{S4}$-consistent set Γ. Let $\mathcal{M}^c$ be the canonical model for $\mathsf{S4}$. By the Truth lemma, we find that $\mathcal{M}^c, \Gamma \Vdash \neg A$, in other words, that $\mathcal{M}^c, \Gamma \nVdash A$. Therefore A is not valid. $\square$

1.3 Embeddings

In this section, we will consider some connections between classical and intuitionistic logic as well as between intuitionistic logic and $\mathsf{S4}$. First, we observe that intuitionistic logic is contained in classical logic.

Lemma 1.48. *For any formula $A \in \mathcal{L}_{ip}$,*

$$\mathsf{IPC} \vdash A \quad \text{implies} \quad \mathsf{CL} \vdash A \ .$$

To establish this lemma, it is enough to show that all instances of **PAx3**–**PAx9** are derivable in CL. Then the claim follows easily by induction on the length of the derivation in IPC.

Note that while A is a formula of $\mathcal{L}_{\mathrm{ip}}$, the statement $\mathsf{CL} \vdash A$ treats it as a classical formula. That means conjunctions and disjunctions are replaced with their corresponding (classically equivalent) implications. Because of this translation, some axiom schemes of IPC are trivially included in some other axiom schemes of CL. For example, the $\mathcal{L}_{\mathrm{ip}}$-formula $Q \to (P \vee Q)$, which is an instance of **PAx7**, becomes $Q \to (\neg P \to Q)$, which is an instance of **PAx1**.

Some other axioms, however, require more work. Let us show that **PAx9** is contained in CL, i.e. $\mathsf{CL} \vdash \bot \to P$. Let G be $\neg\neg P \to P$, which is an instance of **LDN** and hence provable in CL. Further $G \to (\bot \to G)$ is an instance of **PAx1**. Hence by (**MP**) we have $\bot \to G$, which is

$$\bot \to (\neg\neg P \to P) \ . \tag{1.8}$$

The formula $\bot \to ((P \to \bot) \to \bot)$ is an instance of **PAx1**. We can write it as

$$\bot \to \neg\neg P \ . \tag{1.9}$$

1.3 Embeddings

Finally,
$$\bot \to (\neg\neg P \to P) \to (\bot \to \neg\neg P \to (\bot \to P))$$
is an instance of **PAx2**. Hence an application of (**MP**) with (1.8) and another application of (**MP**) with (1.9) yields $\mathsf{CL} \vdash \bot \to P$.

More interesting than Lemma 1.48 is the question whether the converse direction also holds. Of course, not every theorem of CL is a theorem of IPC. In particular **LDN** is not provable in IPC. Still, there might be a sense in which CL is contained in IPC.

Glivenko's theorem, see, e.g., [41, 70, 113] provides such an embedding of CL into IPC.

Theorem 1.49 (Glivenko's theorem). *For any formula $A \in \mathcal{L}_{ip}$,*

$$\mathsf{CL} \vdash A \quad\Longleftrightarrow\quad \mathsf{IPC} \vdash \neg\neg A \ .$$

Glivenko's theorem is not the only embedding we will use. We also need a faithful translation from intuitionistic logic to the modal logic S4.

Definition 1.50. For any formula $A \in \mathcal{L}_{ip}$, we inductively define a formula $\mathsf{Gt}(A) \in \mathcal{L}_\Box$ as follows:

$$\begin{aligned}
\mathsf{Gt}(\bot) &:= \bot \\
\mathsf{Gt}(P) &:= P \\
\mathsf{Gt}(A \vee B) &:= \Box\mathsf{Gt}(A) \vee \Box\mathsf{Gt}(B) \\
\mathsf{Gt}(A \wedge B) &:= \mathsf{Gt}(A) \wedge \mathsf{Gt}(B) \\
\mathsf{Gt}(A \to B) &:= \Box\mathsf{Gt}(A) \to \mathsf{Gt}(B) \ .
\end{aligned}$$

The translation $\mathsf{Gt}(\cdot)$ indeed provides an embedding of IPC into S4. A proof of the following theorem can be found, for instance, in [136].

Theorem 1.51. *For any formula $A \in \mathcal{L}_{ip}$,*

$$\mathsf{IPC} \vdash A \quad\Longleftrightarrow\quad \mathsf{S4} \vdash \mathsf{Gt}(A) \ .$$

In this theorem, the direction from left to right means that the embedding is correct. The converse direction states that the embedding is faithful.

1.4 Notes

Our language for classical propositional logic is based on falsum and implication. Of course, there are many other possible choices for the set of primitive Boolean connectives. Some of these choices—together with their corresponding axiom systems—are discussed in [109]. Historical background on this topic may be found in [43].

Intuitionistic logic has its origin in Brouwer's work on the foundations of mathematics. For a historic account on this topic we refer to [137, 136]. Its connections to computer science are elaborated on in [113].

Modal logic is a very broad subject. A mathematical treatment of modal logic is presented in [28, 41, 79]. Modal logic as epistemic logic is discussed in depth in [50, 107]

There exist several embeddings of IPC into S4. Gödel [72] was the first who introduced an embedding of this kind (see also Section 6.1) and he established its correctness. Later McKinsey and Tarski [108] showed that it also is faithful. The translation Gt(·), which we use here, is taken from Girard's [68] embedding of intuitionistic logic into classical linear logic. Troelstra and Schwichtenberg [136] study Gt(·) as an embedding of IPC into S4 and present proofs for its correctness and faithfulness.

Chapter 2

The Logic of Proofs

The purpose of this chapter is to introduce syntax and semantics for the Logic of Proofs, the first representative of the family of justification logics. We start with defining the language $\mathcal{L}_J$ of justification logic and presenting axioms and rules of the Logic of Proofs. The Logic of Proofs is a justification logic that corresponds to the modal logic S4. This correspondence treats S4 as a logic of provability, i.e., $\Box A$ is read *A is provable* and replaces the provability modality $\Box$ with *justification terms* that represent specific proofs. Thus, the language of justification logic is more precise than that of modal logic: it affords a finer control on the kinds of proof steps comprising the concept of "provable." In fact, under this correspondence, one modal logic, S4, turns into a family of justification logics that differ in the availability of axioms for reasoning. In this section, we show the following properties of the Logic of Proofs:

- the deduction theorem,
- the Logic of Proofs internalizes its own notion of proof provided sufficiently many axioms are available for reasoning (see Definition 2.10 for a formal description of "sufficiently many"), and
- the substitution lemma (carefully formulated).

There are two kinds of models we study. *Basic modular models* are closely related to the syntactic construction of justifications. *Modular models* combine this syntactic construction with relational structures, thus connecting justified knowledge with the traditional notion of knowledge from modal logic. We establish soundness and completeness of the Logic of Proofs for both classes of models. Last but not least we show completeness with respect to a subclass of modular models, called *fully explanatory models*, which require that there be a justification for any formula F that is known in the modal logic sense.

2.1 Syntax

Before we can define the set of formulas for justification logic, we have to specify how we represent justifications and what operations on them we consider. We assume countable sets $\mathsf{JConst} = \{c_0, c_1, \ldots, c_n, \ldots\}$ of *(justification) constants* and $\mathsf{JVar} = \{x_0, x_1, \ldots, x_n, \ldots\}$ of *(justification) variables*.

Definition 2.1 (Justification terms). *Justification terms* of the language $\mathcal{L}_\mathsf{J}$ are defined inductively as follows:

1. each constant $c_i \in \mathsf{JConst}$ and each variable $x_i \in \mathsf{JVar}$ is a justification term;

2. if s and t are terms, then so are

   - $(s \cdot t)$, read *s dot t*,
   - $(s + t)$, read *s plus t*,
   - $!s$, read *bang s*.

Sometimes we call justification term also *evidence terms* and we use these two notions interchangeably. When no confusion is expected, we call them simply *terms*. The set of *atomic terms* is $\mathsf{ATm} := \mathsf{JConst} \cup \mathsf{JVar}$. We denote the *set of all terms* by Tm. A term is called *ground* if it does not contain variables.

Letters r, s, and t, with or without sub- and/or superscripts, are used to denote terms; letters x, y, and z, with or without sub- and/or superscripts, are used to denote justification variables; letters a, b, and c, with or without sub- and/or superscripts, are used to denote justification constants. We use the notation $t(x_{i_1}, \ldots, x_{i_n})$ to emphasize that the term t contains no variables other than $x_{i_1}, \ldots, x_{i_n}$. Given a term $t(x_{i_1}, \ldots, x_{i_k}, \ldots, x_{i_n})$ and a term s, we use $t(x_{i_1}, \ldots, s, \ldots, x_{i_n})$ for the result of replacing every occurrence of x_{i_k} in $t(x_{i_1}, \ldots, x_{i_k}, \ldots, x_{i_n})$ with s.

The unary *proof-checker* operation ! binds stronger than the binary *application* · and *sum* + operations. As for numerical polynomials,

2.1 Syntax

· binds stronger than +, which is why justification terms were originally called *proof polynomials*. Binary operations on terms are left-associative. As usual, parentheses within terms can be omitted if they can be inferred from these rules. For instance,

$$!x \cdot y + a \cdot b \cdot c \quad \text{means} \quad \big((!x \cdot y) + ((a \cdot b) \cdot c)\big).$$

Definition 2.2 (Justification language). We assume the same set of atomic propositions $\mathsf{Prop} = \{P_0, P_1, \ldots, P_n, \ldots\}$ as in $\mathcal{L}_{\mathrm{cp}}$, $\mathcal{L}_{\mathrm{ip}}$, and $\mathcal{L}_{\Box}$. *Formulas* of the language $\mathcal{L}_\mathsf{J}$ are defined inductively as follows:

1. $\bot$ is a formula of $\mathcal{L}_\mathsf{J}$;

2. each atomic proposition $P_i \in \mathsf{Prop}$ is a formula of $\mathcal{L}_\mathsf{J}$;

3. if A and B are formulas of $\mathcal{L}_\mathsf{J}$ and $t \in \mathsf{Tm}$ is a term, then so are

   - $(A \to B)$ and
   - $t{:}A$ read t *justifies* A.

Remark 2.3. Like $\mathcal{L}_\Box$, the language $\mathcal{L}_\mathsf{J}$ is also an extension of $\mathcal{L}_{\mathrm{cp}}$. These two extensions, however, are incomparable.

As is common for unary connectives, $t{:}$ binds more strongly than all binary connectives. The agreements about omitting parentheses for formulas are the same as before.

We define the subformulas of an $\mathcal{L}_\mathsf{J}$-formula as follows:

Definition 2.4 (Subformula). The set of subformulas $\mathsf{sub}(F)$ of an $\mathcal{L}_\mathsf{J}$-formula F is inductively defined by:

1. $\mathsf{sub}(\bot) := \{\bot\}$,

2. $\mathsf{sub}(P_i) := \{P_i\}$ for each $P_i \in \mathsf{Prop}$,

3. $\mathsf{sub}(A \to B) := \{A \to B\} \cup \mathsf{sub}(A) \cup \mathsf{sub}(B)$,

4. $\mathsf{sub}(t{:}A) := \{t{:}A\} \cup \mathsf{sub}(A)$.

Definition 2.5 (Axioms of LP). The *axiom schemes of the Logic of Proofs* are all axiom schemes of CL (see Figure 1.1) and justification axiom schemes from Figure 2.1.

j $x{:}(P \to Q) \to (y{:}P \to x \cdot y{:}Q)$

j+ $x{:}P \vee y{:}P \to (x+y){:}P$

jt $x{:}P \to P$

j4 $x{:}P \to {!}x{:}x{:}P$

Figure 2.1: The justification axiom schemes of the Hilbert-style systems for the Logic of Proofs.

Remark 2.6. Note that we again describe infinitely many axioms by one representing formula, of which they all are substitution instances. The main difference is in the kinds of substitutions employed. In the propositional and modal languages, only substitutions of formulas for atomic propositions were considered. Now for the justification language, we additionally consider substitutions of justification terms for justification variables. For instance, for arbitrary terms t and s and any formula A, we have
$$\left(t{:}A \vee s{:}A \to (t+s){:}A\right) \in \mathbf{j+} \ .$$

Definition 2.7 (Constant specification). A *constant specification* CS for LP is any subset
$$\mathsf{CS} \subseteq \{(c, A) \mid c \in \mathsf{JConst} \text{ and } A \text{ is an axiom of } \mathsf{LP}\} \ .$$
A constant specification is called *finite* if its cardinality is finite.

Constant specifications are used to control which axioms the logic provides justifications for. In particular, the Logic of Proofs is not really one logic, as the modal logic S4. Rather the Logic of Proofs is an umbrella term for a family of logics with the same set of axioms that vary in how many of these axioms have justifications. Such a granularity can be useful, for instance, for modeling agents that are not logically omniscient.

If not stated otherwise, by a *constant specification* we always mean a constant specification for LP.

Definition 2.8 (Logic LP$_{\mathsf{CS}}$). Let CS be a constant specification. The *justification logic* LP$_{\mathsf{CS}}$ is described by a Hilbert-style system with all axioms of LP, i.e., with all axiom schemes of CL (see Figure 1.1) and justification axiom schemes from Figure 2.1 and with the inference rules (**MP**)

2.1 Syntax

(see Figure 1.1) and CS-*axiom necessitation* (**AN**_{CS}):

$$\frac{}{c{:}A} \; (\mathbf{AN_{CS}}), \text{where } (c, A) \in \mathsf{CS} \; .$$

We write $\mathsf{LP_{CS}} \vdash A$ to mean that the *formula A is derivable in* $\mathsf{LP_{CS}}$ and $\Delta \vdash_{\mathsf{LP_{CS}}} A$ to mean that the *formula A is derivable in $\mathsf{LP_{CS}}$ from the set of formulas* Δ. When the logic $\mathsf{LP_{CS}}$ is clear from the context, the subscript $\mathsf{LP_{CS}}$ is omitted. We write Δ, A for $\Delta \cup \{A\}$.

Remark 2.9. Although axiom necessitation is a rule without premises, it is important to consider it as a rule and not as an axiom schema. If we said that $c{:}A$ is an axiom for each $(c, A) \in \mathsf{CS}$, then the notion of an axiom would depend on the notion of a constant specification, which depends on the notion of an axiom. In order to avoid this circularity, we introduce axiom necessitation as a rule.

Definition 2.10 (Axiomatically appropriate constant specification). A constant specification CS is called *axiomatically appropriate* if for each axiom A of LP, there is a constant $c \in \mathsf{JConst}$ such that $(c, A) \in \mathsf{CS}$.

As we will soon see, axiomatic appropriateness is a necessary condition for many important properties of logics of proofs. However, there are many different axiomatically appropriate constant specifications. For instance, it is possible to use one constant to justify all the axioms, a separate constant for each axiom, or to use one constant per axiom scheme. The smallest and largest constant specifications are obvious points of comparison and the corresponding logics of proofs have proper names:

- The Logic of Proofs $\mathsf{LP_{CS}}$ for the smallest constant specification $\mathsf{CS} = \emptyset$ is denoted $\mathsf{LP_0}$.

- The Logic of Proofs $\mathsf{LP_{CS}}$ for the largest constant specification $\mathsf{CS} = \{(c, A) \mid c \in \mathsf{JConst} \text{ and } A \text{ is an axiom of } \mathsf{LP}\}$ is called the Logic of Proofs in the literature and denoted LP.

Most general results[1] regarding Logics of Proofs hold of LP, in part because the largest constant specification is axiomatically appropriate.

Let us briefly discuss how the role of a given justification term depends on its main connective.

- Variables are used to represent arbitrary justifications.

---
[1] With the exception of lower complexity bounds.

- Constants are used to justify assumptions in situations where they are not analyzed any further. In particular, constants are used to justify axioms of the logic (note that all Logics of Proofs LP_{CS} have the same set of axioms). A constant specification CS can be seen as a way to specify which axioms an agent is aware of.

- The operation $\cdot$ represents the agent's ability of applying the rule of modus ponens. Suppose an agent knows A and $A \to B$. By an application of modus ponens, the agent can infer B. Suppose now that t justifies her knowledge of A and s justifies her knowledge of $A \to B$. The inferred knowledge of B can be justified by $s \cdot t$.

- The operation ! represents the agent's ability of performing positive introspection. Suppose an agent knows A and t is a justification for A. By positive introspection, the agent knows that she knows A and that A is justified by t. The evidence term $!t$ represents a justification for the result of the positive introspection act.

- The operation + combines any two justifications to a justification with broader scope. If s is an evidence for A, then whatever evidence t may be, the combined evidence $s + t$ remains evidence for A. One may think of s and t as two volumes of an encyclopedia and $s+t$ as the set of those two volumes. Suppose that one of the volumes, say s, contains justification for a proposition A. Then also the larger set $s + t$ contains justification for A. Actually, this seemingly innocent example has a hidden assumption. What if the second volume contains a corrigendum claiming that A has since been disproved? It turns out that the operation + secretly smuggles *monotonic reasoning* into our logics. Given that the Logics of Proofs was originally developed with formal arithmetical proofs in mind, this assumption is quite reasonable.

Under this earlier arithmetical interpretation of Logics of Proofs, justification terms represent formal proofs and the expression $s{:}A$ means that s represents a proof of A. In this context the operations on justification terms have the following interpretation. The operation $\cdot$ still represents an application of modus ponens. If t is a proof of A and s is a proof of $A \to B$, then $s \cdot t$ is the proof of B that results from s and t by an application of modus ponens. The operation ! represents proof checking. If s is a proof of A, then $!s$ is a proof that s is a proof of A. Finally, for (linear Hilbert-style) proofs s and t, the sum $s + t$ can be

2.1 Syntax

interpreted as a concatenation of s and t. The presence of $+$ implies that evidence terms represent *multi-conclusion* proofs. Indeed if s is a proof of A and t is a proof of B, then $s+t$ is a proof of both A and B.

The interpretation of terms as proofs also explains the **jt** axiom scheme. It states that if x represents a proof of P, then P must be true. That means the **jt** axiom scheme claims soundness of the proofs that are represented by terms.

We now give an example of a derivation in a Logic of Proofs to show how the axioms and the rules of $\mathsf{LP_{CS}}$ work together. It also illustrates the role of the $+$ operation.

Example 2.11. Consider a set Δ such that for terms s and t and for formulas A and B,

$$\Delta := \{s{:}(A \to A \vee B),\ t{:}(B \to A \vee B)\}\ .$$

Then

$$\Delta \vdash_{\mathsf{LP_{CS}}} x{:}A \vee y{:}B \to (s \cdot x + t \cdot y){:}(A \vee B)\ .$$

Proof. Obviously, we have

$$\Delta \vdash_{\mathsf{LP_{CS}}} s{:}(A \to A \vee B) \quad \text{and} \quad \Delta \vdash_{\mathsf{LP_{CS}}} t{:}(B \to A \vee B)\ .$$

Using j and (MP) we obtain

$$\Delta \vdash_{\mathsf{LP_{CS}}} x{:}A \to s \cdot x{:}(A \vee B) \quad \text{and} \quad \Delta \vdash_{\mathsf{LP_{CS}}} y{:}B \to t \cdot y{:}(A \vee B)\ .$$

Finally, from j+, we have

$$\Delta \vdash_{\mathsf{LP_{CS}}} s \cdot x{:}(A \vee B) \to (s \cdot x + t \cdot y){:}(A \vee B)$$

and

$$\Delta \vdash_{\mathsf{LP_{CS}}} t \cdot y{:}(A \vee B) \to (s \cdot x + t \cdot y){:}(A \vee B)\ .$$

Using propositional reasoning, we obtain the desired result. □

Remark 2.12. Single operators $t{:}$ of Logics of Proofs are not normal modalities since they need not satisfy the property

$$t{:}(A \to B) \to (t{:}A \to t{:}B)\ .$$

Thus $\mathsf{LP_{CS}}$ is essentially different from multi-modal logics like, for instance, Propositional Dynamic Logic [52], where each modality is normal and the set of modalities is upgraded by an additional algebraic structure. This contrasts with justification logics, where the $\Box$-modality is decomposed into a family of evidence terms and each of those terms is a non-normal modality.

Since every formula representing an axiom scheme of CL also represents an axiom scheme of $\mathsf{LP_{CS}}$ for any constant specification CS, since $\mathcal{L}_{cp} \subseteq \mathcal{L}_J$, and since modus ponens, the only inference rule of CL, is also an inference rule of $\mathsf{LP_{CS}}$, each logic $\mathsf{LP_{CS}}$ is an *extension* of CL. Indeed, we have the following theorem:

Theorem 2.13 (Conservativity of $\mathsf{LP_{CS}}$ over CL). *For any constant specification* CS, *the logic* $\mathsf{LP_{CS}}$ *is a conservative extension of* CL, *i.e., for any formula* $A \in \mathcal{L}_{cp}$,

$$\mathsf{LP_{CS}} \vdash A \quad \Longleftrightarrow \quad \mathsf{CL} \vdash A \ .$$

Proof. The proof follows along the lines of Theorem 1.29. Again it is sufficient to show the left-to-right implication. The translation we need should erase all terms in a given formula. Consider the translation $(\cdot)^s \colon \mathcal{L}_J \to \mathcal{L}_{cp}$ defined by induction on the structure of the modal formula:

1. $P^s := P$;

2. $\bot^s := \bot$;

3. $(B \to C)^s := B^s \to C^s$;

4. $(t{:}B)^s := B^s$.

By induction on the length of an $\mathsf{LP_{CS}}$-derivation of $A \in \mathcal{L}_J$ we prove that $\mathsf{CL} \vdash A^s$ whenever $\mathsf{LP_{CS}} \vdash A$ and note that $A^s = A$ for any $A \in \mathcal{L}_{cp}$. The cases for propositional axioms and for (**MP**) have been covered in the proof of Theorem 1.29. Only new cases are discussed below:

1. If A is an instance of the **j**, **jt**, or **j4** axiom scheme, then A^s is a classical propositional tautology of the form $C \to C$.

2. If A is an instance of the **j+** axiom scheme, then A^s is a classical propositional tautology of the form $C \vee C \to C$.

2.2 Basic Properties

3. If $A = c{:}B$ is obtained by ($\mathbf{AN_{CS}}$), where $(c, B) \in \mathsf{CS}$ and B is an axiom of LP, then $A^s = B^s$ is an s-translation of an axiom of LP, which we have demonstrated to be a classical propositional tautology. □

We use the same conventions regarding propositional reasoning in the justification logic $\mathsf{LP_{CS}}$ as in $\mathsf{S4}$.

2.2 Basic Properties

We start with a simple observation concerning constant specifications.

Lemma 2.14. *For any constant specification* CS *and any* $\mathcal{L}_\mathsf{J}$*-formula* F *with* $\mathsf{LP_{CS}} \vdash F$ *we have*

1. $\mathsf{LP_{CS'}} \vdash F$ *for any constant specification* $\mathsf{CS'} \supseteq \mathsf{CS}$;

2. $\mathsf{LP_{CS'}} \vdash F$ *for some finite constant specification* $\mathsf{CS'} \subseteq \mathsf{CS}$.

Proof. The first claim is trivial. The second claim follows immediately from the fact that any $\mathsf{LP_{CS}}$-proof of F may contain only finitely many applications of the axiom necessitation rule. □

Corollary 2.15. *For any constant specification* CS *and any formula* F *of* $\mathcal{L}_\mathsf{J}$, *we have*

$$\mathsf{LP_0} \vdash F \quad \Longrightarrow \quad \mathsf{LP_{CS}} \vdash F \quad \Longrightarrow \quad \mathsf{LP} \vdash F \ .$$

In other words, $\mathsf{LP_0}$ *is the smallest Logic of Proofs and* LP *is the largest one.*

2.2.1 Deduction Theorem

The deduction theorem is standard for justification logics.

Theorem 2.16 (Deduction Theorem)**.** *For an arbitrary constant specification* CS, *for any set* Δ *of* $\mathcal{L}_\mathsf{J}$*-formulas, and for any* $\mathcal{L}_\mathsf{J}$*-formulas* A *and* B,

$$\Delta, A \vdash_{\mathsf{LP_{CS}}} B \quad \Longleftrightarrow \quad \Delta \vdash_{\mathsf{LP_{CS}}} A \to B \ .$$

Proof. The direction from right to left is trivial. Assume that we have

$$\Delta \vdash_{\mathsf{LP_{CS}}} A \to B \ .$$

Then also $\Delta, A \vdash_{\mathsf{LP}_{\mathsf{CS}}} A \to B$. Clearly, we also have $\Delta, A \vdash_{\mathsf{LP}_{\mathsf{CS}}} A$. Thus we conclude by modus ponens $\Delta, A \vdash_{\mathsf{LP}_{\mathsf{CS}}} B$.

The direction from left to right is shown by induction on the length of the derivation $\Delta, A \vdash_{\mathsf{LP}_{\mathsf{CS}}} B$. We distinguish the following cases:

1. If B is an axiom of LP, or $B \in \Delta$, or B is obtained by ($\mathbf{AN_{CS}}$), then $\Delta \vdash_{\mathsf{LP}_{\mathsf{CS}}} B$. Since $B \to (A \to B)$ is an instance of the axiom scheme **PAx1**, we have $\Delta \vdash_{\mathsf{LP}_{\mathsf{CS}}} B \to (A \to B)$. By modus ponens we conclude $\Delta \vdash_{\mathsf{LP}_{\mathsf{CS}}} A \to B$.

2. If $B = A$, then $A \to B$ is a substitution instance of a classical propositional tautology. Therefore, $\Delta \vdash_{\mathsf{LP}_{\mathsf{CS}}} A \to B$.

3. If the formula B is inferred by (**MP**) from C and $C \to B$, then by the induction hypothesis,
$$\Delta \vdash_{\mathsf{LP}_{\mathsf{CS}}} A \to C \quad \text{and} \quad \Delta \vdash_{\mathsf{LP}_{\mathsf{CS}}} A \to (C \to B) \ .$$
Since $A \to (C \to B) \to (A \to C \to (A \to B))$ is an instance of the axiom scheme **PAx2**, we find by modus ponens
$$\Delta \vdash_{\mathsf{LP}_{\mathsf{CS}}} A \to C \to (A \to B) \ .$$
Another application of modus ponens yields $\Delta \vdash_{\mathsf{LP}_{\mathsf{CS}}} A \to B$. □

Remark 2.17. The Deduction Theorem is well known to hold for both classical and intuitionistic propositional logics. However, when formulated in a naive manner, it fails for modal logics, including S4. More precisely, if the modal necessitation rule is allowed to be applied to all formulas previously derived, then $P \vdash_{\mathsf{S4}} \Box P$, whereas $\mathsf{S4} \nvdash P \to \Box P$. In order to make the Deduction Theorem work in the modal environment, the definition of derivation from hypotheses, $\Delta \vdash B$, must be changed: namely, the use of (**Nec**) must be restricted to formulas derivable without any hypotheses.

2.2.2 Internalization Property

An important property of justification logics is their ability to internalize their own notion of proof, as stated in the following lemma.

Lemma 2.18 (Internalization for Variables). *Let CS be an axiomatically appropriate constant specification. For arbitrary formulas $A, B_1, \ldots, B_n$ of $\mathcal{L}_\mathsf{J}$, if*
$$B_1, \ldots, B_n \vdash_{\mathsf{LP}_{\mathsf{CS}}} A \ ,$$

2.2 Basic Properties

then there is a term $t(y_1, \ldots, y_n) \in \mathsf{Tm}$ such that

$$y_1{:}B_1, \ldots, y_n{:}B_n \vdash_{\mathsf{LP_{CS}}} t(y_1, \ldots, y_n){:}A$$

for arbitrary variables $y_1, \ldots, y_n$.

Proof. By induction on the length of the derivation $B_1, \ldots, B_n \vdash_{\mathsf{LP_{CS}}} A$. We distinguish the following cases:

1. If A is an axiom of LP, then since CS is axiomatically appropriate, there exists a constant c with $(c, A) \in \mathsf{CS}$ and we set $t := c$. Then $y_1{:}B_1, \ldots, y_n{:}B_n \vdash_{\mathsf{LP_{CS}}} c{:}A$ by axiom necessitation and c contains no variables.

2. If $A = B_i$, we set $t := y_i$. Then $y_1{:}B_1, \ldots, y_n{:}B_n \vdash_{\mathsf{LP_{CS}}} y_i{:}B_i$ follows trivially and y_i contains no variables other than y_i.

3. If A is inferred by (**MP**) from C and $C \to A$, then by the induction hypothesis, there are terms r and s containing no variables other than $y_1, \ldots, y_n$ such that

$$y_1{:}B_1, \ldots, y_n{:}B_n \vdash_{\mathsf{LP_{CS}}} r{:}C$$

and

$$y_1{:}B_1, \ldots, y_n{:}B_n \vdash_{\mathsf{LP_{CS}}} s{:}(C \to A) \ .$$

Since

$$s{:}(C \to A) \to (r{:}C \to s \cdot r{:}A)$$

in an instance of j, we conclude by propositional reasoning that

$$y_1{:}B_1, \ldots, y_n{:}B_n \vdash_{\mathsf{LP_{CS}}} s \cdot r{:}A \ .$$

Since any variable occurring in $s \cdot r$ must occur either in s or in r, it is sufficient to set $t := s \cdot r$.

4. If A is obtained by ($\mathbf{AN_{CS}}$), then $A = c{:}F$ for some constant c. Since

$$c{:}F \to !c{:}c{:}F$$

is an instance of **j4**, we conclude by propositional reasoning that

$$y_1{:}B_1, \ldots, y_n{:}B_n \vdash_{\mathsf{LP_{CS}}} !c{:}c{:}F \ .$$

Since $!c$ contains no variables, it is sufficient to set $t := !c$. □

Note that in the above lemma, the justification term $t(y_1, \ldots, y_n)$, which justifies A, is an exact blueprint of the given $\mathsf{LP}_{\mathsf{CS}}$-derivation of A. While any set of variables $y_1, \ldots, y_n$ can be used, it is often reasonable to choose fresh variables, i.e., variables that have not yet been used in any way. If the given derivation employs no hypotheses, that is if $n = 0$, then the resulting statement is called *constructive necessitation*, which essentially is a justification counterpart of the modal necessitation rule.

Corollary 2.19 (Constructive Necessitation). *Let* CS *be an axiomatically appropriate constant specification. For any formula* $A \in \mathcal{L}_\mathsf{J}$, *if* $\mathsf{LP}_{\mathsf{CS}} \vdash A$, *then* $\mathsf{LP}_{\mathsf{CS}} \vdash t{:}A$ *for some ground term* $t \in \mathsf{Tm}$.

Combining the previous results, we also obtain internalization for the case when the derivation hypotheses are justified by arbitrary terms.

Corollary 2.20 (Internalization for Arbitrary Terms). *Let* CS *be an axiomatically appropriate constant specification. For arbitrary* $\mathcal{L}_\mathsf{J}$-*formulas* $A, B_1, \ldots, B_n$ *and arbitrary terms* $s_1, \ldots, s_n \in \mathsf{Tm}$, *if*

$$B_1, \ldots, B_n \vdash_{\mathsf{LP}_{\mathsf{CS}}} A ,$$

then there is a term $t \in \mathsf{Tm}$ *such that*

$$s_1{:}B_1, \ldots, s_n{:}B_n \vdash_{\mathsf{LP}_{\mathsf{CS}}} t{:}A .$$

Proof. Assume $B_1, \ldots, B_n \vdash_{\mathsf{LP}_{\mathsf{CS}}} A$. Applying Deduction Theorem 2.16 several times, we get

$$\mathsf{LP}_{\mathsf{CS}} \vdash B_1 \to (B_2 \to \cdots \to (B_n \to A)\cdots) .$$

By Constructive Necessitation, there is a ground term s' such that

$$\mathsf{LP}_{\mathsf{CS}} \vdash s'{:}\Big(B_1 \to (B_2 \to \cdots \to (B_n \to A)\cdots)\Big) .$$

Using several instances of the axiom scheme **j** and propositional reasoning, we obtain

$$s_1{:}B_1, \ldots, s_n{:}B_n \vdash_{\mathsf{LP}_{\mathsf{CS}}} s' \cdot s_1 \cdot s_2 \cdots s_n{:}A .$$

Thus, it is sufficient to set $t := s' \cdot s_1 \cdot s_2 \cdots s_n$. □

Remark 2.21. While we do not yet have enough tools to demonstrate it, the requirement that the constant specification be axiomatically appropriate in all three preceding statements is necessary in the general case. Indeed, if A is an axiom that has no constant to justify it and if A cannot be derived from the other axioms of LP, then it is possible to show that $\mathsf{LP}_{\mathsf{CS}} \vdash A$ but $\mathsf{LP}_{\mathsf{CS}} \nvdash t{:}A$ for any term t.

2.2 Basic Properties

2.2.3 Substitution Property

As mentioned earlier in addition to substituting formulas for atomic propositions, in justification logics arbitrary justification terms can be substituted for justification variables. However, any substitution may necessitate an adjustment to the constant specification if derivability is to be preserved.

Example 2.22. Consider the constant specification

$$\mathsf{CS} := \{(c, x{:}P \to P)\} \ .$$

By axiom necessitation, we have $\mathsf{LP}_{\mathsf{CS}} \vdash c{:}(x{:}P \to P)$. Suppose now that we replace the atomic proposition P with some arbitrary formula F different from P. It can be shown (using some additional machinery) that

$$\mathsf{LP}_{\mathsf{CS}} \nvdash c{:}(x{:}F \to F)$$

because the pair $(c, x{:}F \to F)$ is not an element of the constant specification CS. In order to restore the derivability, the same substitution needs to be applied to the constant specification: for

$$\mathsf{CS}' := \mathsf{CS}[P \mapsto F] = \{(c, x{:}F \to F)\} \ ,$$

we have $\mathsf{LP}_{\mathsf{CS}'} \vdash c{:}(x{:}F \to F)$.

The same phenomenon occurs when we substitute an arbitrary term t for the variable x.

We have already used substitutions in defining axioms of LP. Now we provide the formal description and the notation for this concept:

Definition 2.23 (Substitution). A *substitution* is a mapping σ from justification variables to justification terms and from atomic propositions to formulas. Substitutions are then extended to terms and formulas in the obvious way: for an $\mathcal{L}_\mathsf{J}$-formula B, the formula $B\sigma$ is obtained from B by simultaneously replacing all occurrences of x_i with $\sigma(x_i)$ and all occurrences of P_i with $\sigma(P_i)$ in B for all $i \geq 0$. For a set of formulas Φ, we define $\Phi\sigma := \{B\sigma \mid B \in \Phi\}$ and for a constant specification CS, we define $\mathsf{CS}\sigma := \{(c, A\sigma) \mid (c, A) \in \mathsf{CS}\}$.

Definition 2.24 (Substitution composition). Let σ and τ be substitutions. Their composition $\sigma \circ \tau$ (understood as the application of τ first and the application of σ to the result) is defined as follows:

$$(\sigma \circ \tau)(x_i) := \bigl(\tau(x_i)\bigr)\sigma \;,$$
$$(\sigma \circ \tau)(P_i) := \bigl(\tau(P_i)\bigr)\sigma \;.$$

The following lemma trivially follows from the definitions:

Lemma 2.25. *Let σ and τ be substitutions. For any term $t \in \mathsf{Tm}$ and any formula $A \in \mathcal{L}_\mathsf{J}$ we have*

$$t(\sigma \circ \tau) = (t\tau)\sigma \;,$$
$$A(\sigma \circ \tau) = (A\tau)\sigma \;.$$

In other words, any consecutive application of several substitutions can be viewed as a single substitution.

The following trivial lemma states the correctness of the substitution operation with respect to constant specifications.

Lemma 2.26. *If CS is a constant specification for LP, so is $\mathsf{CS}\sigma$ for any substitution σ.*

Proof. We need to prove that $A\sigma$ is an axiom of LP whenever A is. According to our definition, each axiom of LP is a substitution instance of a formula representing one of the axioms schemes, i.e., in our new notation, $A = F\tau$ for some formula F from Figure 1.1 or Figure 2.1 and some substitution τ. Using the preceding lemma,

$$A\sigma = (F\tau)\sigma = F(\sigma \circ \tau)$$

is a substitution instance of the same axiom scheme and, hence, is an axiom. □

Lemma 2.27 (Substitution). *For any constant specification CS, any set of formulas $\Delta \subseteq \mathcal{L}_\mathsf{J}$, any formula $A \in \mathcal{L}_\mathsf{J}$, and any substitution σ,*

$$\Delta \vdash_{\mathsf{LP}_\mathsf{CS}} A \quad \text{implies} \quad \Delta\sigma \vdash_{\mathsf{LP}_{\mathsf{CS}\sigma}} A\sigma \;.$$

Proof. By induction on the length of the derivation $\Delta \vdash_{\mathsf{LP}_\mathsf{CS}} A$. We distinguish the following cases:

2.2 Basic Properties

1. If A is an axiom of LP, then, as discussed earlier, $A\sigma$ is also an axiom of LP.

2. If $A \in \Delta$, then $A\sigma \in \Delta\sigma$ and, thus, $\Delta\sigma \vdash_{\mathsf{LP}_{\mathsf{CS}\sigma}} A\sigma$.

3. If A is inferred by (**MP**) from C and $C \to A$, then, by the induction hypothesis,

$$\Delta\sigma \vdash_{\mathsf{LP}_{\mathsf{CS}\sigma}} C\sigma \quad \text{and} \quad \Delta\sigma \vdash_{\mathsf{LP}_{\mathsf{CS}\sigma}} (C \to A)\sigma \ .$$

Given that $(C \to A)\sigma = C\sigma \to A\sigma$, by (**MP**) we get

$$\Delta\sigma \vdash_{\mathsf{LP}_{\mathsf{CS}\sigma}} A\sigma \ .$$

4. If A is obtained by ($\mathbf{AN_{CS}}$), then $A = c{:}F$ with $(c, F) \in \mathsf{CS}$. Thus, $(c, F\sigma) \in \mathsf{CS}\sigma$ and, by ($\mathbf{AN_{CS\sigma}}$) we get

$$\Delta\sigma \vdash_{\mathsf{LP}_{\mathsf{CS}\sigma}} c{:}(F\sigma) \ ,$$

which is the same as

$$\Delta\sigma \vdash_{\mathsf{LP}_{\mathsf{CS}\sigma}} (c{:}F)\sigma \ . \qquad \square$$

Definition 2.28 (Schematic constant specification). We say that a constant specification CS is *schematic* if it satisfies the following property: for each constant c_i, the set of axioms $\{A \mid (c_i, A) \in \mathsf{CS}\}$ consists of all instances of one or several (possibly zero) axiom schemes of LP.

Schematicness is important when we deal with substitutions for the following reason. Let σ be a substitution. If A is an instance of a given axiom scheme, then, as shown in the proof of Lemma 2.26, $A\sigma$ is an instance of the same scheme. Hence, if CS is schematic and $(c, A) \in \mathsf{CS}$, then we also have $(c, A\sigma) \in \mathsf{CS}$. That is schematic constant specifications are closed under substitutions. Thus, we get the following corollary of Lemma 2.27.

Corollary 2.29 (Substitution Property). *Let CS be a schematic constant specification. We have for any set of formulas $\Delta \subseteq \mathcal{L}_\mathsf{J}$, any $\mathcal{L}_\mathsf{J}$-formula A and any substitution σ*

$$\Delta \vdash_{\mathsf{LP}_{\mathsf{CS}}} A \quad \text{implies} \quad \Delta\sigma \vdash_{\mathsf{LP}_{\mathsf{CS}}} A\sigma \ .$$

Remark 2.30. The substitution property above is well known to hold without restrictions for both classical and intuitionistic propositional logics and for many modal logics, including S4 with respect to substitutions of formulas for atomic propositions.

Remark 2.31. Replacement is another property that holds for both classical and intuitionistic propositional logics and for many modal logics. However, this property does not and should not hold for justification logics. A logic L has the *replacement property* if for any formula F of the underlying language with exactly one occurrence of an atomic proposition P and for arbitrary formulas A and B of the same language

$$\mathsf{L} \vdash A \leftrightarrow B \quad \text{implies} \quad \mathsf{L} \vdash F(A) \leftrightarrow F(B) \;,$$

where $F(C)$ is an abbreviation of $F[P \mapsto C]$ for any formula C.

Many modal logics, including S4, enjoy the replacement property. Justification logics, however, do not satisfy it. Consider two axioms A and B of LP and let $\mathsf{CS} := \{(c, A)\}$. Obviously, we have $\mathsf{LP_{CS}} \vdash A \leftrightarrow B$ and $\mathsf{LP_{CS}} \vdash c{:}A$. However, if we replace A with B in $c{:}A$, it can be shown that we no longer have a provable formula: $\mathsf{LP_{CS}} \nvdash c{:}B$ since $(c, B) \notin \mathsf{CS}$.

In fact, we would not want the replacement property to hold by design. Justification terms are to present explicit reasons for provability/knowledge of certain statements. The replacement property suggests that any subformula can be replaced with a provably equivalent formula. However, such a replacement is bound to affect the justification terms present in the formula subjected to the replacement. In particular, the proof of the equivalence between the formula being replaced and the formula replacing it must be incorporated into these justification terms. In addition, such a replacement with justification-term recomputation can only be applied to implications, not to equivalences. The reason for that is that equivalence $\leftrightarrow$ does not respect polarities of formulas. A precise formulation of the replacement property suitable for justification logics and an algorithm to compute justification terms in the formula after the replacement is presented in [54].

2.3 Basic Modular Models

We now give semantics for justifications logics. We start with symbolic models in this section and present epistemic, that is Kripke-style, models in the next section.

2.3 Basic Modular Models

The essential question when giving semantics for justification logics is: *what is a justification?* We will address this issue only within a precise mathematical theory where the question has the form: *what is the logical type of justifications?* In (basic) modular models, justifications are interpreted as *sets of formulas*. We regard a formula as being justified by a term t if and only if it lies in the set that interprets the term t. The semantics for justification logics has to respect the operations on evidence terms. Thus we need corresponding operations on sets of formulas, which are defined next.

Definition 2.32. Let $X, Y \subseteq \mathcal{L}_J$ and $t \in \mathsf{Tm}$. We define

$$X \cdot Y := \{F \in \mathcal{L}_J \mid G \to F \in X \text{ and } G \in Y$$
$$\text{for some formula } G \in \mathcal{L}_J\};$$
$$t{:}X := \{t{:}F \mid F \in X\}.$$

Informally, $X \cdot Y$ is the result of applying modus ponens to all members of X and Y in this order. The set $t{:}X$ is the result of prefixing all members of X with the operator $t{:}$.

We retain a classical interpretation of propositions as truth values. In addition, the interpretation of an evidence term t is a set of formulas t^*. The intended meaning of $F \in t^*$ is that t *justifies* F. Thus, while propositions are interpreted semantically, justifications are interpreted syntactically. This is an important feature: a model may treat distinct formulas F and G as equivalent (both are evaluated to true or both are evaluated to false), but still be able to distinguish the justification assertions $t{:}F$ and $t{:}G$. That is we may have semantically equivalent formulas F and G such that $F \in t^*$ but $G \notin t^*$.

Definition 2.33 (Basic evaluation). Let CS be an arbitrary constant specification. A *basic evaluation for* $\mathsf{LP}_{\mathsf{CS}}$, or a basic $\mathsf{LP}_{\mathsf{CS}}$-evaluation, is a function $*$ that maps atomic propositions to truth values $0, 1$ and maps justification terms to sets of formulas:

$$*\colon \mathsf{Prop} \to \{0,1\} \quad \text{and} \quad *\colon \mathsf{Tm} \to \mathcal{P}(\mathcal{L}_J) \;,$$

such that for arbitrary $s, t \in \mathsf{Tm}$ and any $F \in \mathcal{L}_J$,

(1) $s^* \cdot t^* \subseteq (s \cdot t)^*$;

(2) $s^* \cup t^* \subseteq (s + t)^*$;

(3) $F \in t^*$ if $(t, F) \in \mathsf{CS}$;

(4) $s{:}(s^*) \subseteq (!s)^*$.

Here P^* for $P \in \mathsf{Prop}$ and t^* for $t \in \mathsf{Tm}$ denote $*(P)$ and $*(t)$, respectively.

Definition 2.34 (Truth under a basic evaluation). We define what it means for a formula to *hold under a basic evaluation* $*$ inductively as follows:

- $* \not\Vdash \bot$;

- $* \Vdash P$ if and only if $P^* = 1$ for any $P \in \mathsf{Prop}$;

- $* \Vdash F \to G$ if and only if $* \not\Vdash F$ or $* \Vdash G$;

- $* \Vdash t{:}F$ if and only if $F \in t^*$.

A basic evaluation is a trivial thing. We assign arbitrary truth values to atomic propositions as usual. Then, essentially we regard formulas $t{:}A$ as atomic as assign them truth values as well. However, we must do so in such a way as to respect the axioms. The **j**, **j+**, and **j4** axiom schemes tell us that various formulas $t{:}A$ should be made true if other related formulas are true, which is easily achieved in a basic evaluation.

The situation is different for the **jt** axiom scheme. In a basic evaluation, we may have the situation that a term t justifies a formula A where A is evaluated to false. We have to eliminate this possibility since we are looking for models of $\mathsf{LP}_{\mathsf{CS}}$, which includes all formulas $t{:}A \to A$ as axioms.

Definition 2.35 (Factive evaluation). Let CS be an arbitrary constant specification. A basic $\mathsf{LP}_{\mathsf{CS}}$-evaluation $*$ is called *factive* if $F \in t^*$ implies $* \Vdash F$ for all $t \in \mathsf{Tm}$ and $F \in \mathcal{L}_\mathsf{J}$.

Definition 2.36 (Basic modular model). Let CS be an arbitrary constant specification. A *basic modular model for* $\mathsf{LP}_{\mathsf{CS}}$, or a basic modular $\mathsf{LP}_{\mathsf{CS}}$-model, is a basic $\mathsf{LP}_{\mathsf{CS}}$-evaluation $*$ that is factive.

Remark 2.37. Note that we have not yet established that any basic modular model exists. We can show this only later in Corollary 2.44.

We immediately find that $\mathsf{LP}_{\mathsf{CS}}$ is sound with respect to basic modular $\mathsf{LP}_{\mathsf{CS}}$-models.

2.3 Basic Modular Models

Theorem 2.38 (Soundness: Basic modular models). *For any constant specification* CS *and any formula* F *of* $\mathcal{L}_\mathsf{J}$ *we have*

$$\mathsf{LP}_{\mathsf{CS}} \vdash F \quad \Longrightarrow \quad * \Vdash F \text{ for all basic modular } \mathsf{LP}_{\mathsf{CS}}\text{-models } *.$$

Proof. As usual, the proof is by induction on the length of the derivation of F. Let $*$ be a basic modular $\mathsf{LP}_{\mathsf{CS}}$-model. Since basic modular models extend propositional truth tables, it is obvious that all instances of propositional axiom schemes hold under $*$ and that the rule (**MP**) is respected by the semantics.

1. If $F = s{:}(A \to B) \to (t{:}A \to s \cdot t{:}B)$ is an instance of **j**, assume that $* \Vdash s{:}(A \to B)$ and $* \Vdash t{:}A$. It means that $A \to B \in s^*$ and $A \in t^*$. Consequently, $B \in s^* \cdot t^*$ by the definition of $\cdot$ on sets. By the definition of basic evaluations, $B \in (s \cdot t)^*$, which means that $* \Vdash s \cdot t{:}B$.

2. If $F = s{:}A \lor t{:}A \to (s+t){:}A$ is an instance of **j+**, the proof is similar and we omit it.

3. If $F = t{:}A \to A$ is an instance of **jt**, assume that $* \Vdash t{:}A$. It means that $A \in t^*$, so $* \Vdash A$ by factivity of $*$.

4. If $F = t{:}A \to !t{:}t{:}A$ is an instance of **j4**, assume that $* \Vdash t{:}A$. It means that $A \in t^*$. Consequently, $t{:}A \in t{:}(t^*)$ by the definition of $t{:}$ on sets. By the definition of basic evaluations, $t{:}A \in (!t)^*$, which means that $* \Vdash !t{:}t{:}A$.

5. If F is obtained by (**AN**$_\mathsf{CS}$), then $F = c{:}A$ for some $(c, A) \in \mathsf{CS}$. Thus, $A \in c^*$ by the definition of basic evaluations for $\mathsf{LP}_{\mathsf{CS}}$. We conclude that $* \Vdash c{:}A$. □

An important question to be asked about any new theory is whether it is non-trivial, i.e., whether there are statements it does not endorse.

Definition 2.39 (Consistent theory). We call a logical theory L *consistent* if the empty set of formulas is L-consistent.

Since the conjunction over the empty set is $\top$, we find by Definition 1.40 that a logical theory L is consistent if and only if

$$\mathsf{L} \nvdash \top \to \bot,$$

which means $\mathsf{L} \nvdash \bot$.

Remark 2.40. For a logical theory that is given in a language $\mathcal{L}$ and that is based on classical (or intuitionistic) propositional logic, such as LP$_{CS}$ or S4, we can equivalently formulate consistency as

$$\{A \mid \mathsf{L} \vdash A\} \neq \mathcal{L}$$

because in such theories $\mathsf{L} \vdash \bot \to A$ for any formula $A \in \mathcal{L}$.

It is common to prove consistency of a theory by demonstrating that it has models. In the case of classical propositional logic CL, the existence of models is obvious: any assignment of truth values to atomic propositions defines a truth table for CL. Since $\bot$ is false in any truth table, it cannot be a theorem of CL by soundness. Thus, CL is consistent.

The argument for S4 and for intuitionistic propositional logic is exactly the same except that the appropriate type of models needs to be used. For instance, for S4, one might consider a model consisting of a singleton reflexive world supplied with an arbitrary valuation function.

The situation for LP$_{CS}$ is not that simple. The problem is that, unlike in the case of CL, it is not immediately obvious that models exist. A basic modular model for LP$_{CS}$ has to satisfy the conditions of a basic evaluation and it has to be factive. While satisfying the properties of a basic evaluation is reasonably simple (for instance, one might define t^* to be $\mathcal{L}_J$ for each term t), ensuring factivity is not so easy. Hence, consistency of LP$_{CS}$ is not an immediate consequence of soundness with respect to basic modular models.[2] However, we can reduce the consistency of LP$_{CS}$ to the consistency of CL. And from that we can deduce the existence of basic modular models.

Lemma 2.41 (Consistency of LP$_{CS}$). *For any constant specification* CS, *the logic* LP$_{CS}$ *is consistent.*

Proof. Assume towards a contradiction that LP$_{CS}$ were not consistent, that means LP$_{CS} \vdash \bot$. By the conservativity of LP$_{CS}$ over CL (Theorem 2.13), it would then follow that CL $\vdash \bot$, which is not the case. □

Since LP$_{CS}$ is consistent, we know that the empty set is an LP$_{CS}$-consistent set of formulas. Given the Lindenbaum Lemma, this implies that there exists a maximal LP$_{CS}$-consistent set. The following lemma

---

[2] There are other semantics for LP$_{CS}$, e.g., M-models [114] and F-models [53], however, where a model can be constructed as a least fixed point of a monotone operator, making consistency an easy corollary.

2.3 Basic Modular Models

shows that any maximal LP$_{CS}$-consistent set induces a factive basic LP$_{CS}$-evaluation. Hence we can conclude that there exists a basic modular model for LP$_{CS}$.

Lemma 2.42. *Let CS be an arbitrary constant specification. For any maximal LP$_{CS}$-consistent set Φ the function $*_\Phi$ defined by*

$$P^{*\Phi} := \begin{cases} 1 & P \in \Phi \\ 0 & P \notin \Phi \end{cases} \quad \text{and} \quad t^{*\Phi} := \{B \in \mathcal{L}_J \mid t{:}B \in \Phi\} \quad (2.1)$$

for any $P \in \mathsf{Prop}$ and any $t \in \mathsf{Tm}$ is a basic modular LP$_{CS}$-model, i.e., a factive basic LP$_{CS}$-evaluation.

Proof. Since the argument is uniform for all maximal consistent sets Φ, we fix one such Φ for the duration of this proof and write $*$ instead of $*_\Phi$ (except in the formulation of the Truth Lemma below). Let us first check that $*$ is a basic LP$_{CS}$-evaluation.

Condition (1). Suppose $A \in s^* \cdot t^*$. Then there is $B \in \mathcal{L}_J$ such that $B \to A \in s^*$ and $B \in t^*$. By (2.1), $s{:}(B \to A) \in \Phi$ and $t{:}B \in \Phi$. Since

$$\mathsf{LP_{CS}} \vdash s{:}(B \to A) \to (t{:}B \to s \cdot t{:}A)$$

is an instance of **j**, by Lemma 1.42.1 for the maximal consistent set Φ,

$$s{:}(B \to A) \to (t{:}B \to s \cdot t{:}A) \in \Phi \ .$$

Using Lemma 1.42.4 twice for the maximal consistent set Φ, we conclude that $s \cdot t{:}A \in \Phi$. Thus, $A \in (s \cdot t)^*$ by (2.1).

Condition (2). Suppose $A \in s_1^* \cup s_2^*$, i.e., $A \in s_i^*$ for some $i = 1, 2$. By (2.1), $s_i{:}A \in \Phi$. Since

$$\mathsf{LP_{CS}} \vdash s_i{:}A \to s_1{:}A \vee s_2{:}A \quad \text{and} \quad \mathsf{LP_{CS}} \vdash s_1{:}A \vee s_2{:}A \to (s_1{+}s_2){:}A$$

by propositional reasoning and as an instance of **j+** respectively, it follows by Lemma 1.42.1

$$s_i{:}A \to s_1{:}A \vee s_2{:}A \in \Phi \quad \text{and} \quad s_1{:}A \vee s_2{:}A \to (s_1 + s_2){:}A \in \Phi \ .$$

Now, by Lemma 1.42.4 twice, we conclude that $(s_1 + s_2){:}A \in \Phi$ and, hence, by (2.1), $A \in (s_1 + s_2)^*$.

Condition (3). Suppose $(t, F) \in \mathsf{CS}$. Then $\mathsf{LP}_{\mathsf{CS}} \vdash t{:}F$ by axiom necessitation. Thus, $t{:}F \in \Phi$ by Lemma 1.42.1 and $F \in t^*$ by (2.1).

Condition (4). Suppose $A \in s{:}(s^*)$. Then $A = s{:}B$ for some $B \in s^*$. By (2.1), $s{:}B \in \Phi$. Since $\mathsf{LP}_{\mathsf{CS}} \vdash s{:}B \to {!}s{:}s{:}B$ as an instance of **j4**, it follows by Lemma 1.42.1 $s{:}B \to {!}s{:}s{:}B \in \Phi$. Now, by Lemma 1.42.4, we conclude that ${!}s{:}s{:}B \in \Phi$ and, hence, by (2.1), $s{:}B \in ({!}s)^*$, i.e., $A \in ({!}s)^*$.

We still need to show that $*$ is factive. But we cannot do it until the *Truth Lemma* is proved:

Lemma 2.43 (Truth Lemma). *Let CS be an arbitrary constant specification. Let Φ be a maximal $\mathsf{LP}_{\mathsf{CS}}$-consistent set. Then, for the basic $\mathsf{LP}_{\mathsf{CS}}$-evaluation $*_\Phi$ defined by (2.1) the following property holds: for all $D \in \mathcal{L}_\mathsf{J}$,*

$$D \in \Phi \quad \Longleftrightarrow \quad *_\Phi \Vdash D \ . \tag{2.2}$$

Proof. We establish (2.2) by induction on the structure of D:

1. If $D = P \in \mathsf{Prop}$, then $\quad P \in \Phi \ \Leftrightarrow\ P^* = 1 \ \Leftrightarrow\ * \Vdash P$.

2. If $D = \bot$ or $D = A \to B$, the proof is the same as in Lemma 1.46.

3. If $D = t{:}B$, then $\quad t{:}B \in \Phi \ \Leftrightarrow\ B \in t^* \ \Leftrightarrow\ * \Vdash t{:}B$. $\square$

Finally, we show that $*$ is a basic modular $\mathsf{LP}_{\mathsf{CS}}$-model by demonstrating the factivity of $*$. In doing so and in all future proofs we will stop providing details regarding which part of Lemma 1.42 is used, employing "maximal consistent set" as a reference to any statement from Lemma 1.42. Suppose $B \in t^*$. Then $t{:}B \in \Phi$ by (2.1). Since $t{:}B \to B$ is an instance of **jt** and Φ is a maximal consistent set, we have $B \in \Phi$. Now $* \Vdash B$ follows by (2.2). $\square$

Corollary 2.44 (Model existence). *For any constant specification CS, the logic $\mathsf{LP}_{\mathsf{CS}}$ has a basic modular model.*

Proof. By Lemma 2.41 we know that $\mathsf{LP}_{\mathsf{CS}}$ is consistent, that is $\emptyset$ is an $\mathsf{LP}_{\mathsf{CS}}$-consistent set. By the Lindenbaum Lemma we find a maximal $\mathsf{LP}_{\mathsf{CS}}$-consistent set $\Phi \supseteq \emptyset$. By Lemma 2.42 we know that $*_\Phi$ as defined by (2.1) is a basic modular model for $\mathsf{LP}_{\mathsf{CS}}$. $\square$

Using the Lindenbaum Lemma, Lemma 2.42, and the Truth Lemma it is easy to show completeness of $\mathsf{LP}_{\mathsf{CS}}$.

Theorem 2.45 (Completeness: Basic modular models). *For any constant specification CS and any $F \in \mathcal{L}_\mathsf{J}$ we have*

$$*\Vdash F \text{ for all basic modular } \mathsf{LP}_{\mathsf{CS}}\text{-models } * \quad \Longrightarrow \quad \mathsf{LP}_{\mathsf{CS}} \vdash F \ .$$

Proof. Assume that $\mathsf{LP}_{\mathsf{CS}} \nvdash F$. Then $\{\neg F\}$ is $\mathsf{LP}_{\mathsf{CS}}$-consistent and, by Lindenbaum Lemma 1.41, is contained in some maximal $\mathsf{LP}_{\mathsf{CS}}$-consistent set $\Phi \subseteq \mathcal{L}_\mathsf{J}$. Since $\neg F \in \Phi$, we find $*_\Phi \Vdash \neg F$ by (2.2) for the basic modular model $*_\Phi$ constructed from Φ according to Lemma 2.42. Completeness of $\mathsf{LP}_{\mathsf{CS}}$ follows by contraposition. □

Remark 2.46. The case for $t{:}B$ in the proof of Truth Lemma 2.43 for justification logics is much simpler than the corresponding case for $\Box B$ in the proof of Truth Lemma 1.46 for modal logics. The former does not even rely on the induction hypothesis.

2.4 Epistemic Models

The main purpose of epistemic models, also known as possible world models or Kripke-style models, is to connect justification logic to traditional epistemic logics, which rely heavily on possible world semantics. To define epistemic models for $\mathsf{LP}_{\mathsf{CS}}$ we start with a Kripke frame and assign to each possible world a basic $\mathsf{LP}_{\mathsf{CS}}$-evaluation. However, this is not enough since these basic evaluations may have nothing to do with the epistemic structure of the model that is represented by the accessibility relation. The notions of justification (in the sense of the basic evaluations) and of knowledge (in the sense of the Kripke frame) will be connected by the following principle:

having a specific justification for F must yield knowing that F.
(JYB)

This principle is called *justification yields belief* since it applies not only to $\mathsf{LP}_{\mathsf{CS}}$ but also to weaker justification logics that formalize notions of belief rather than notions of knowledge.

Definition 2.47 (Quasimodel). We define a *quasimodel for* $\mathsf{LP}_{\mathsf{CS}}$, or an $\mathsf{LP}_{\mathsf{CS}}$-quasimodel, to be a triple $\mathcal{M} = (W, R, *)$, where $W \neq \varnothing$ is a non-empty set of (possible) worlds, $R \subseteq W \times W$ is an accessibility relation, and the *evaluation* $*$ maps each world $w \in W$ to a basic $\mathsf{LP_{CS}}$-evaluation $*_w$. For quasimodels, we will write P_w^* instead of $*_w(P)$ or P^{*w} and t_w^* instead of $*_w(t)$ or t^{*w}.

Definition 2.48 (Truth in quasimodels). We define what it means for a formula A to *hold at a world* $w \in W$ *of a quasimodel* $\mathcal{M} = (W, R, *)$, written $\mathcal{M}, w \Vdash A$, inductively as follows:

$\mathcal{M}, w \Vdash P$ if and only if $P_w^* = 1$ for $P \in \mathsf{Prop}$;

$\mathcal{M}, w \nVdash \bot$;

$\mathcal{M}, w \Vdash F \to G$ if and only if $\mathcal{M}, w \nVdash F$ or $\mathcal{M}, w \Vdash G$;

$\mathcal{M}, w \Vdash t{:}F$ if and only if $F \in t_w^*$.

We say that a formula A is *valid* in a model $\mathcal{M}$ and write $\mathcal{M} \Vdash A$ if $\mathcal{M}, w \Vdash A$ for all $w \in W$.

For a given quasimodel $\mathcal{M} = (W, R, *)$ and a world $w \in W$, we define

$$\Box_w := \{F \in \mathcal{L}_\mathsf{J} \mid \mathcal{M}, v \Vdash F \text{ whenever } R(w,v)\} \; . \qquad (2.3)$$

Looking back at Kripke models for $\mathsf{S4}$, it is easy to see that $F \in \Box_w$ means[3] that $\Box F$ holds at the world w of the model, i.e., that F is known in w. This observation facilitates the following compact formulation of the JYB principle:

$$t_w^* \subseteq \Box_w \quad \text{for all } t \in \mathsf{Tm} \text{ and } w \in W \; . \qquad \text{(JYB)}$$

Since the accessibility relation R plays no role in the definition of truth in quasimodels, it is easy to see that truth at a world w of a quasimodel $\mathcal{M} = (W, R, *)$ coincides with truth at the corresponding basic evaluation $*_w$:

Lemma 2.49 (Locality of truth in quasimodels). *Let $\mathcal{M} = (W, R, *)$ be an arbitrary $\mathsf{LP_{CS}}$-quasimodel. For any world $w \in W$ and any formula $G \in \mathcal{L}_\mathsf{J}$,*

$$\mathcal{M}, w \Vdash G \quad \Longleftrightarrow \quad *_w \Vdash G \; . \qquad (2.4)$$

By analogy with basic modular models, we define the notion of factivity for quasimodels.

---
[3]Modulo the different language we are using.

2.4 Epistemic Models

Definition 2.50 (Factive quasimodel). A quasimodel $\mathcal{M} = (W, R, *)$ is called *factive* if $F \in t_w^*$ implies $\mathcal{M}, w \Vdash F$ for all $w \in W$, $t \in \mathsf{Tm}$, and $F \in \mathcal{L}_\mathsf{J}$.

Definition 2.51 (Modular model). A *modular model for* LP_CS, or a modular LP_CS-model, is an LP_CS-quasimodel $\mathcal{M} = (W, R, *)$ that meets the following conditions:

(1) $t_w^* \subseteq \Box_w$ for all $t \in \mathsf{Tm}$ and $w \in W$; \hfill (JYB)

(2) R is reflexive;

(3) R is transitive.

These conditions may seem superfluous since R plays no role in determining the truth of formulas. But Conditions (2) and (3) are well known for the corresponding modal logic $\mathsf{S4}$ and, hence, are needed, so to say, for backward compatibility: they ensure that the same semantics can be used for justification logics and modal logics. Condition (1) plays, in this respect, the role of a catalyzer allowing for a transition between the two formalisms by ensuring that modal belief exist whenever a justification is available.

Note that, unlike for the case of basic modular models, we do not require that modular models be factive. Instead, this property is derived from JYB and the reflexivity restriction on R. Hence JYB together with reflexivity generalizes factivity.

The transitivity restriction is, strictly speaking, not needed if we are only interested in the logics of proofs. It was shown by Antonakos [3] that LP_CS remains sound if Condition (3) is dropped from the definition of modular models. However, in the next chapter we are going to study the relationship of LP_CS and $\mathsf{S4}$. There it is important (see the proof of Lemma 3.13) that the canonical model for LP_CS is also a model of $\mathsf{S4}$, which means that R should be transitive.

Lemma 2.52 (Modular models are factive). *All modular* LP_CS*-models* $\mathcal{M} = (W, R, *)$ *are factive.*

Proof. Whenever $F \in t_w^*$ for some $F \in \mathcal{L}_\mathsf{J}$, some $t \in \mathsf{Tm}$, and some $w \in W$, we have $F \in \Box_w$ by JYB. Since $R(w, w)$ by the reflexivity of R, we obtain $\mathcal{M}, w \Vdash F$ from (2.3). $\square$

Corollary 2.53 (Factivity of basic evaluations used in modular models)**.** *For any modular* $\mathsf{LP}_{\mathsf{CS}}$*-model* $\mathcal{M} = (W, R, *)$ *and any world* $w \in W$, *the basic evaluation* $*_w$ *is factive and, hence, a basic modular model for* $\mathsf{LP}_{\mathsf{CS}}$.

Proof. Assume that for the basic evaluation $*_w$, we have $A \in t^{*w}$ for some formula $A \in \mathcal{L}_\mathsf{J}$ and some term $t \in \mathsf{Tm}$. Then $A \in t_w^*$ in the modular model notation. By Lemma 2.52, we get $\mathcal{M}, w \Vdash A$, from which we conclude $*_w \Vdash A$ by (2.4). □

There is an additional property that follows from JYB but is peculiar to the possible-worlds scenario.

Lemma 2.54 (Monotonicity)**.** *Let* $\mathcal{M} = (W, R, *)$ *be a modular* $\mathsf{LP}_{\mathsf{CS}}$*-model. Then for any* $t \in \mathsf{Tm}$ *and for arbitrary* $w, v \in W$, *if* $R(w, v)$, *then* $t_w^* \subseteq t_v^*$, *i.e., justifications remain relevant in accessible worlds.*

Proof. Assume $R(w, v)$ and $F \in t_w^*$ for some formula $F \in \mathcal{L}_\mathsf{J}$. Then we have $t{:}F \in (!t)_w^*$ because $*_w$ is a basic evaluation for $\mathsf{LP}_{\mathsf{CS}}$. So $t{:}F \in \square_w$ by JYB and, in particular, $\mathcal{M}, v \Vdash t{:}F$ by (2.3), which means that $F \in t_v^*$. □

The soundness and completeness of justification logics with respect to modular models are almost obvious:

Theorem 2.55 (Soundness and completeness: Modular models)**.** *For any constant specification* CS *and any* $F \in \mathcal{L}_\mathsf{J}$ *we have*

$$\mathsf{LP}_{\mathsf{CS}} \vdash F \quad \Longleftrightarrow \quad \mathcal{M} \Vdash F \text{ for all modular } \mathsf{LP}_{\mathsf{CS}}\text{-models } \mathcal{M} \ . \tag{2.5}$$

Proof. It is sufficient to prove that any formula refutable by a basic modular model can be refuted at a world in a modular model and vice versa.

Soundness. Let $\mathcal{M} = (W, R, *)$ be a modular $\mathsf{LP}_{\mathsf{CS}}$-model. We need to show that any formula $F \in \mathcal{L}_\mathsf{J}$ such that $\mathsf{LP}_{\mathsf{CS}} \vdash F$ holds at any world $w \in W$. By Corollary 2.53, we know that $*_w$ is a basic modular $\mathsf{LP}_{\mathsf{CS}}$-model. By soundness of $\mathsf{LP}_{\mathsf{CS}}$ with respect to basic modular models, we get $*_w \Vdash F$. Hence, $\mathcal{M}, w \Vdash F$ by (2.4)

Completeness. For the opposite direction, suppose $\mathsf{LP}_{\mathsf{CS}} \nvdash F$. By completeness of $\mathsf{LP}_{\mathsf{CS}}$ with respect to basic modular models, there exists a basic modular $\mathsf{LP}_{\mathsf{CS}}$-model $*$ such that $* \nVdash F$. We define an $\mathsf{LP}_{\mathsf{CS}}$-quasimodel $\mathcal{M} := (\{1\}, R, \star)$ with $R := \{(1, 1)\}$ and $\star_1 := *$. Then

2.4 Epistemic Models

by (2.4) we have $\mathcal{M}, 1 \not\Vdash F$ and it only remains to show that $\mathcal{M}$ is a modular $\mathsf{LP_{CS}}$-model, i.e., that all the restrictions on R and the condition JYB are met. The reflexivity and transitivity of R are trivial. Let us finish the proof by demonstrating JYB. Assume that $G \in t_1^*$ for some formula $G \in \mathcal{L}_\mathsf{J}$, some term $t \in \mathsf{Tm}$, and the only world 1. Then $G \in t^*$ by the definition of $\star$. Since $*$ is factive, $* \Vdash G$. Thus, $\mathcal{M}, 1 \Vdash G$ by (2.4) and, consequently, $G \in \square_1$. □

It is reasonable to ask whether our semantics supports that every belief is evidenced by some justification, in other words, whether for each formula F

if F holds in all accessible worlds, then there is a justification for F.

Definition 2.56 (Fully explanatory modular models). A modular $\mathsf{LP_{CS}}$-model $\mathcal{M} = (W, R, *)$ is *fully explanatory* if for any $w \in W$ and any $F \in \mathcal{L}_\mathsf{J}$, $F \in \square_w$ implies $F \in t_w^*$ for some $t \in \mathsf{Tm}$.

This notion can be seen as the converse of JYB and, taking the latter into account, can be reformulated as $\square_w = \bigcup_{t \in \mathsf{Tm}} t_w^*$. We show soundness and completeness of $\mathsf{LP_{CS}}$ with respect to fully explanatory models using an adaptation of the canonical model construction. Note, however, that this strengthening of the completeness theorem requires constant specification to be axiomatically appropriate.

Theorem 2.57 (Soundness and completeness: Fully explanatory modular models). *Let CS be an axiomatically appropriate constant specification. Then $\mathsf{LP_{CS}}$ is sound and complete with respect to fully explanatory modular $\mathsf{LP_{CS}}$-models.*

Proof. Soundness immediately follows from soundness with respect to all modular $\mathsf{LP_{CS}}$-models (and holds independent of whether CS is axiomatically appropriate).

The hard task is to show completeness. In Lemma 2.42, we showed how to construct a basic modular model $*_\Phi$ from any maximal $\mathsf{LP_{CS}}$-consistent set Φ. From all such $*_\Phi$'s we create the canonical modular $\mathsf{LP_{CS}}$-model and use (2.4) to transfer the properties of $*_\Phi$ demonstrated as part of the completeness proof for basic modular models. Thus, we define $\mathcal{M}^c := (W^c, R^c, *^c)$, where

$$W^c := \{\Phi \subseteq \mathcal{L}_\mathsf{J} \mid \Phi \text{ is a maximal } \mathsf{LP_{CS}}\text{-consistent set}\}$$

and $*_\Phi^c$ for each $\Phi \in W$ is defined by (2.1). Finally, we set

$$R^c(\Phi, \Psi) \quad \text{iff} \quad \Phi/\sharp \subseteq \Psi \ ,$$

where

$$\Phi/\sharp := \{F \in \mathcal{L}_\mathsf{J} \mid t{:}F \in \Phi \text{ for some term } t\} \ .$$

To show that $\mathcal{M}^c$ is a modular $\mathsf{LP_{CS}}$-model, it remains to establish that the set W^c is non-empty, that R^c is reflexive and transitive, and that the condition JYB is satisfied.

We start with showing $W^c \neq \emptyset$. From the proof of Corollary 2.44 we know that the empty set is $\mathsf{LP_{CS}}$-consistent. Thus, by the Lindenbaum lemma, there exists a maximal $\mathsf{LP_{CS}}$-consistent set extending $\emptyset$, which is an element of W^c.

To show JYB, assume $F \in t_\Phi^{*c}$ for some $F \in \mathcal{L}_\mathsf{J}$, some $t \in \mathsf{Tm}$, and some $\Phi \in W^c$. We need to show that $F \in \square_\Phi$, i.e., that $\mathcal{M}^c, \Psi \Vdash F$ whenever $R^c(\Phi, \Psi)$. Consider any such $\Psi \in W^c$. We have $t{:}F \in \Phi$ by the definition of $*_\Phi^c$ and $F \in \Psi$ by the definition of R^c. By (2.2), $*_\Psi^c \Vdash F$, and, finally, $\mathcal{M}^c, \Psi \Vdash F$ follows from (2.4).

To show that R^c is reflexive, consider any $\Phi \in W^c$. Assume that $F \in \Phi/\sharp$, i.e., that $t{:}F \in \Phi$ for some $t \in \mathsf{Tm}$. By maximal consistency of Φ and since $t{:}F \to F$ is an axiom of $\mathsf{LP_{CS}}$ we conclude $F \in \Phi$. Therefore, $\Phi/\sharp \subseteq \Phi$, which means that $R^c(\Phi, \Phi)$.

To show that R^c is transitive, consider arbitrary $\Phi, \Psi, \Xi \in W^c$ such that $R^c(\Phi, \Psi)$ and $R^c(\Psi, \Xi)$. Assume that $F \in \Phi/\sharp$, i.e., that $t{:}F \in \Phi$ for some $t \in \mathsf{Tm}$. By maximal consistency of Φ and since $t{:}F \to !t{:}t{:}F$ is an axiom of $\mathsf{LP_{CS}}$ we conclude $!t{:}t{:}F \in \Phi$. Hence $t{:}F \in \Phi/\sharp \subseteq \Psi$ and $F \in \Psi/\sharp \subseteq \Xi$. Therefore, $\Phi/\sharp \subseteq \Xi$, which means $R^c(\Phi, \Xi)$.

Finally, we show that $\mathcal{M}^c$ is fully explanatory. Assume that $F \in \square_\Phi$ for some $F \in \mathcal{L}_\mathsf{J}$ and $\Phi \in W^c$. Then

$$\Phi/\sharp \cup \{\neg F\} \text{ is not } \mathsf{LP_{CS}}\text{-consistent.} \qquad (2.6)$$

Suppose towards a contradiction that (2.6) does not hold. Then, by the Lindenbaum lemma, there would exist a maximal $\mathsf{LP_{CS}}$-consistent set Ψ such that $\Psi \supseteq \Phi/\sharp \cup \{\neg F\}$ and, by the definition of R^c, we would have $R^c(\Phi, \Psi)$. By the maximal consistency of Ψ, it would follow that $F \notin \Psi$. We would have $*_\Psi^c \nVdash F$ by (2.2), and $\mathcal{M}^c, \Psi \nVdash F$ by (2.4), which would contradict our assumption that $F \in \square_\Phi$.

The now established (2.6) implies that there exists a finite set

$$\Sigma = \{G_1, \ldots, G_n\} \subseteq \Phi/\sharp$$

for some $n \geq 0$ such that
$$\mathsf{LP_{CS}} \vdash \neg F \wedge \bigwedge \Sigma \to \bot \ .$$
By propositional reasoning, this implies
$$G_1, \ldots, G_n \vdash_{\mathsf{LP_{CS}}} F \ .$$
Since each $G_i \in \Phi/\sharp$, there must exist terms $s_i \in \mathsf{Tm}$ such that $s_i{:}G_i \in \Phi$ for each $1 \leq i \leq n$. By Corollary 2.20, given the axiomatic appropriateness of CS, there exists a term t such that
$$s_1{:}G_1, \ldots, s_n{:}G_n \vdash_{\mathsf{LP_{CS}}} t{:}F \ .$$
By Deduction Theorem 2.16,
$$\mathsf{LP_{CS}} \vdash s_1{:}G_1 \to (s_2{:}G_2 \to \cdots \to (s_n{:}G_n \to t{:}F) \cdots) \ ,$$
which implies $t{:}F \in \Phi$ by the maximal consistency of Φ. Thus, $F \in t_\Phi^{*^c}$. We have established that $F \in t_\Phi^{*^c}$ for some $t \in \mathsf{Tm}$ whenever $F \in \Box_\Phi$, meaning that $\mathcal{M}^c$ is fully explanatory. $\square$

Remark 2.58. We noted earlier in Remark 2.46 that the proof of the Truth Lemma for $\mathsf{LP_{CS}}$ is much simpler than for $\mathsf{S4}$. However, when we impose the condition on models to be fully explanatory, the complexity returns and, indeed, the preceding proof that the canonical modular $\mathsf{LP_{CS}}$-model is fully explanatory is largely isomorphic to the proof of the more difficult half of the Truth Lemma for the canonical $\mathsf{S4}$-model for formulas of the type $\Box B$.

2.5 Notes

The axiomatic system $\mathsf{LP_{CS}}$ and its basic properties like the deduction theorem, internalization, and the substitution lemma already appear in Artemov's first papers about the Logic of Proofs [4, 6]. Since Artemov aimed at a provability semantics for the modal logic $\mathsf{S4}$, the original semantics for $\mathsf{LP_{CS}}$ was an arithmetical interpretation where justification terms are interpreted as codes for proofs in Peano arithmetic. We will present this approach in Chapter 6.

Mkrtychev[114] introduced the first non-arithmetical semantics for justification logics in order to obtain decidability results for $\mathsf{LP_{CS}}$. Like basic modular models, his models, which are called M-models, are symbolic models. They consist of an evaluation function $*$ that assigns to each evidence term t the set of formulas t^* that are justified by t. Remember that basic modular models only consider factive evaluation functions and use the truth definition

$$t \text{ is evidence for } F \text{ is true} \quad \text{if and only} \quad \text{if } F \in t^*. \qquad (2.7)$$

M-models, in contrast, allow any evaluation function. However, in order to deal with factivity of $\mathsf{LP_{CS}}$, M-models use the truth definition

$$t \text{ is evidence for } F \text{ is true} \quad \text{if and only} \quad \text{if } F \in t^* \text{ and } F \text{ is true.}$$

The clarity of (2.7) is muddled in M-models by truth of F thrown into the mix. This was necessary in order to obtain the desired decidability results. Mkrtychev made a pragmatic choice of efficiency over philosophical transparency.

Later, Fitting [53], working independently from Mkrtychev, presented the first possible world semantics for justification logic, which is commonly referred to as F-models. He essentially developed the same machinery for handling evidence terms as in M-models and combines it with traditional Kripke-style semantics. Hence he uses the following truth definition: a formula $t{:}F$ holds at a world w if and only if

1. F is evidenced by t at world w and
2. F is true at all worlds that are reachable from w.

F-models proved to be very successful for many applications. Still they present the same compromise as M-models for $\mathsf{LP_{CS}}$. For the sake of efficiency, justification and truth are intertwined in that t is only evidence for F if F is true (in the same M-model or at all accessible worlds in an F-model). The philosophical objections to such a paradigm also have practical roots. In court, evidence is used to determine the truth of the matter. However, if the acceptability of the evidence were to depend on this truth, it would create a vicious circle.

A clear ontological separation between evidence and truth is achieved in (basic) modular models. They are based on (2.7), which means that the truth of F does not matter for the truth of $t{:}F$. The connection between the interpretation of justifications and the traditional possible world semantics is achieved by the principle of justification yields belief, which does not affect the truth definition.

2.5 Notes

Artemov [10] introduced modular models for the justification counterpart of the basic modal logic K. Based on his work, Kuznets and Studer [99] studied modular models for $\mathsf{LP}_{\mathsf{CS}}$ and all other justification logics that correspond to a logic from the modal cube, see [61].

Recently, Lehmann and Studer [103] presented subset model for justification logic. Their main idea is to interpret justification terms as sets of possible worlds. A formula $t{:}F$ is true in a subset model if F holds at each world of the interpretation of t, i.e., the interpretation of t is a subset of the interpretation of F.

Some notes on terminology for constant specifications. The definitions of *constant specification* in [5, 6] and of *axiom specification* in [4] correspond to that we call here a *finite constant specification*. The definition that we use here was perhaps first presented in [114]. The term *axiomatically appropriate* is due to Fitting [53]. The term *schematic* was first introduced by Milnikel [110] although its idea goes back to [90, 114].

In the early days of justification logic, there was hope that Gabbay's labelled deductive systems [60] may serve as a natural framework for the Logic of Proofs, see [6]. Unfortunately, this has not been fulfilled. Although there are some similarities, it turned out that the two approaches do not interact well [14].

Chapter 3

Relations with S4

There is a close relationship between the modal logic S4 and justification logics $\mathsf{LP}_{\mathsf{CS}}$. Any evidential assertion $t{:}F$ has a natural modal counterpart $\Box F$ where $\Box F$ can be read as *for some x we have $x{:}F$*. This observation leads us to the notion of the *forgetful projection* $F° \in \mathcal{L}_\Box$ of a justification formula $F \in \mathcal{L}_\mathsf{J}$. The modal formula $F°$ is the result of replacing all occurrences of $t{:}$ in F with $\Box$. Obviously, different justification logic formulas may have the same forgetful projection. In other words, by transforming F into $F°$ we lose some information, which explains the name forgetful projection.

It is easy to see that if A is an axiom of LP, then $A°$ is a provable formula of S4. Also the rules of $\mathsf{LP}_{\mathsf{CS}}$ translate to rules of S4. For the case of axiom necessitation, consider a pair (c, A) of CS. Then A is an axiom of $\mathsf{LP}_{\mathsf{CS}}$ and thus $A°$ is a provable formula of S4. By necessitation we get $\mathsf{S4} \vdash \Box A°$, which is $\mathsf{S4} \vdash (c{:}A)°$. Hence we obtain for any constant specification CS and for any $\mathcal{L}_\mathsf{J}$-formula F that

$$\mathsf{LP}_{\mathsf{CS}} \vdash F \quad \text{implies} \quad \mathsf{S4} \vdash F° \, ,$$

which is established formally below in Lemma 3.2.

There is also a converse result, which, however, requires a much more complicated proof and restrictions on allowable types of constant specifications. Assume that CS is a schematic and axiomatically appropriate constant specification. The realization theorem then states that for each $\mathcal{L}_\Box$-formula F,

$$\mathsf{S4} \vdash F \quad \text{implies} \quad \mathsf{LP}_{\mathsf{CS}} \vdash F^r$$

for some $\mathcal{L}_\mathsf{J}$-formula F^r such that $(F^r)° = F$. Such a formula F^r is called a *realization* of F since each $\Box$-operator of F is realized by some evidence term in F^r.

We present two proofs of the realization theorem. First, in Section 3.2, we show a semantic proof. It relies on the fact that modular models provide a semantics not only for the Logics of Proofs but also for the modal logic S4 and that in fully explanatory modular models, we have that if $\Box F$ holds at a world w in the sense of S4, then there is a term t such that $t{:}F$ holds at w in the sense of $\mathsf{LP_{CS}}$. This semantic proof only establishes the existence of the realization F^r.

In Section 3.3, we provide a construction of a realization F^r, which is based on a cut-free sequent calculus for S4. We start with a cut-free proof of the $\mathcal{L}_\Box$-formula F and show how to build F^r based on the given cut-free proof. The main ingredient for this construction is a generalization of the internalization property, the so-called Lifting Lemma, which makes it possible to 'realize' the ($\Box \supset$)-rule of the sequent calculus for S4.

3.1 Forgetful Projection

We define the forgetful projection F° of an $\mathcal{L}_\mathsf{J}$-formula F by induction on the structure of F and establish that F° is provable in S4 whenever F is provable in $\mathsf{LP_{CS}}$.

Definition 3.1 (Forgetful projection). The mapping $\circ : \mathcal{L}_\mathsf{J} \to \mathcal{L}_\Box$ is defined as follows:

1. $P^\circ := P$ for $P \in \mathsf{Prop}$;

2. $\bot^\circ := \bot$;

3. $(A \to B)^\circ := A^\circ \to B^\circ$;

4. $(t{:}A)^\circ := \Box A^\circ$.

Since we use the same defined Boolean connectives in $\mathcal{L}_\mathsf{J}$ and $\mathcal{L}_\Box$, it is clear that $(A * B)^\circ = A^\circ * B^\circ$ for each binary Boolean connective $*$ and that $(\neg A)^\circ = \neg A^\circ$.

Lemma 3.2. *For any constant specification* CS *and any formula F of $\mathcal{L}_\mathsf{J}$,*

$$\mathsf{LP_{CS}} \vdash F \quad \text{implies} \quad \mathsf{S4} \vdash F^\circ \ .$$

Proof. The proof is by induction on the length of the LP$_{CS}$-derivation of F.

1. If F is an $\mathcal{L}_J$-instance of an axiom scheme from Figure 1.1, then F° is an $\mathcal{L}_\Box$-instance of the same axiom scheme and, therefore, S4 $\vdash F^\circ$.

2. If $F = t{:}(A \to B) \to (s{:}A \to t \cdot s{:}B)$ is an instance of **j**, then
$$F^\circ = \Box(A^\circ \to B^\circ) \to (\Box A^\circ \to \Box B^\circ) \ ,$$
which is an instance of **k**.

3. If $F = s{:}A \lor t{:}A \to (s+t){:}A$ is an instance of **j+**, then
$$F^\circ = \Box A^\circ \lor \Box A^\circ \to \Box A^\circ \ ,$$
which is a classical propositional tautology. Hence, S4 $\vdash F^\circ$.

4. If $F = t{:}A \to A$ is an instance of **jt**, then $F^\circ = \Box A^\circ \to A^\circ$, which is an instance of **t**.

5. If $F = t{:}A \to !t{:}t{:}A$ is an instance of **j4**, then $F^\circ = \Box A^\circ \to \Box\Box A^\circ$, which is an instance of **4**.

6. If F is inferred by (**MP**) from premises A and $A \to F$, then by the induction hypothesis, S4 $\vdash A^\circ$ and S4 $\vdash (A \to F)^\circ$. Since, by definition, $(A \to F)^\circ = A^\circ \to F^\circ$, we can apply (**MP**) and conclude S4 $\vdash F^\circ$.

7. If F is obtained by (**AN**$_{CS}$), then F has the form $c{:}A$ where A is an axiom of LP. Thus, as shown above in 1–5, S4 $\vdash A^\circ$. By (**Nec**) we get S4 $\vdash \Box A^\circ$, i.e., S4 $\vdash F^\circ$. (Note that the exact contents of CS are irrelevant for this argument.) □

3.2 Semantic Realization

We established Lemma 3.2 by a simple induction on the length of the LP$_{CS}$-derivation. Showing the converse, that every S4-theorem is the forgetful projection of some theorem of LP$_{CS}$, requires more involved methods. In the following we present a semantic proof of this property.

Definition 3.3 (Realization). A *realization* is a mapping $r\colon \mathcal{L}_\Box \to \mathcal{L}_J$ such that $(r(A))^\circ = A$ for each $\mathcal{L}_\Box$-formula A.

We can think of a realization as a function that replaces each occurrence of $\Box$ in an $\mathcal{L}_\Box$-formula with some evidence term, thus creating a formula of $\mathcal{L}_J$.

Definition 3.4. We say that a justification logic LP_CS *realizes* $\mathsf{S4}$ if there is a realization r such that for any $\mathcal{L}_\Box$-formula A we have

$$\mathsf{S4} \vdash A \quad \text{implies} \quad \mathsf{LP}_\mathsf{CS} \vdash r(A) \ .$$

Remark 3.5. Given an $\mathsf{S4}$-theorem, there may be essentially different realizations for it in LP_CS. For instance,

$$\Box F \vee \Box F \to \Box F$$

has the realizations

$$x{:}F \vee y{:}F \to x+y{:}F \quad \text{and} \quad x{:}F \vee x{:}F \to x{:}F \ .$$

The former of these realizations is a specification of the $+$-operator whereas the latter one is a trivial tautology.

Remark 3.6. Of course, not every LP_CS realizes $\mathsf{S4}$. In particular, for the empty constant specification $\mathsf{CS} = \varnothing$ it can be shown that $\mathsf{LP}_\varnothing \nvdash t{:}A$ for any term t and any formula A. As a result, no modal formula $\Box B$ is a forgetful projection of a theorem of $\mathsf{LP}_\varnothing$. In order for a justification logic to realize $\mathsf{S4}$, the former needs to be able to emulate the modal necessitation rule by constructive necessitation (Corollary 2.19). As mentioned in Remark 2.21, axiomatic appropriateness of the constant specification is necessary for that.

The aim of this section is to establish a realization theorem, i.e., to identify properties of a constant specification CS that are sufficient to ensure that LP_CS realizes $\mathsf{S4}$.

For the rest of this section, we fix an arbitrary modal formula A of $\mathcal{L}_\Box$ and assume that CS is an axiomatically appropriate constant specification. Let $\mathcal{M}^c = (W^c, R^c, *^c)$ be the fully explanatory canonical model from the proof of Theorem 2.57. We formulate two properties of $\mathcal{M}^c$ that will be important for establishing a realization theorem:

Lemma 3.7 (Truth Lemma for $\mathcal{M}^c$). *For any $\Phi \in W^c$ and any $D \in \mathcal{L}_J$,*

$$D \in \Phi \quad \Longleftrightarrow \quad \mathcal{M}^c, \Phi \Vdash D \ . \tag{3.1}$$

3.2 Semantic Realization

Proof. Since each $\Phi \in W^c$ is a maximal $\mathsf{LP_{CS}}$-consistent set, we have $D \in \Phi$ iff $*_\Phi \Vdash D$ (by Truth Lemma 2.43) iff $*_\Phi^c \Vdash D$ (by the definition of $*^c$). We know from Lemma 2.42 that $*_\Phi^c = *_\Phi$ is a basic $\mathsf{LP_{CS}}$-evaluation for each $\Phi \in W^c$, meaning that $\mathcal{M}^c$ an $\mathsf{LP_{CS}}$-quasimodel.[1] Thus, we can apply Lemma 2.49 to conclude that

$$*_\Phi^c \Vdash D \quad \text{iff} \quad \mathcal{M}^c, \Phi \Vdash D \ .$$

$\square$

Lemma 3.8. *Let* CS *be an axiomatically appropriate constant specification. For any* $\Phi \in W^c$ *and any* $D \in \mathcal{L}_\mathsf{J}$,

if $\Phi/\sharp \cup \{\neg D\}$ *is not* $\mathsf{LP_{CS}}$-*consistent,*

then there is a term t *such that* $t{:}D \in \Phi$. (3.2)

Proof. This implication for axiomatically appropriate CS was proved as part of the proof of Theorem 2.57. $\square$

Definition 3.9 (Polarity). To each subformula occurrence within our fixed modal formula A we assign the positive or negative *polarity*.

1. To the only occurrence of A within A we assign the positive polarity.

2. If a polarity is already assigned to an occurrence of $B \to C$ within A, then the same polarity is assigned to the corresponding occurrence of C and the opposite polarity is assigned to the corresponding occurrence of B.

3. If a polarity is already assigned to an occurrence of $\Box B$ within A, then the same polarity is assigned to the corresponding subformula occurrence of B.

We introduce a notation for singling out occurrences of the same subformula within A. For instance, if $A = B \to B$, we need a compact notation to be able to distinguish the occurrence of B in the antecedent of the implication from the occurrence of the same formula B in the consequent. The notation we suggest may not be the most elegant one, but our goal is to keep it compact. We will denote occurrences of B within A by $o(B)$, $o_1(B)$, $o_2(B)$, etc. Polarity of such occurrences is

---

[1] In Theorem 2.57 we actually proved a stronger statement that $\mathcal{M}^c$ is a fully explanatory modular $\mathsf{LP_{CS}}$-model.

denoted by a superscript $+$ or $-$. Whenever we use $o_i^+(B \to C)$, $o_i^+(C)$, and $o_i^-(B)$ within the same sentence/paragraph, these occurrences of B and C are the immediate suboccurrences of this occurrence of $B \to C$. Similarly for $o_i^-(B \to C)$, $o_i^-(C)$, and $o_i^+(B)$; for $o_i^+(\Box B)$ and $o_i^+(B)$; and for $o_i^-(\Box B)$ and $o_i^-(B)$. Thus, we use indices i for two purposes: to distinguish between different occurrences of the same formula and to connect an occurrence of a formula with its immediate suboccurrence.

Definition 3.10 (Annotation). An *annotation* (for our fixed modal formula A) is an injective mapping $\mathcal{A}$ from the set

$$\{o^-(\Box B) \mid o^-(\Box B) \text{ is a negative occurrence of } \Box B \text{ within } A\}$$

to the set

$$\mathsf{JVar} \ .$$

For the rest of this section, we additionally fix an arbitrary annotation $\mathcal{A}$ for A. We now define a mapping that assigns a set of *potential pre-realizations* to each subformula occurrence within A with respect to our fixed annotation $\mathcal{A}$. It is important that different occurrences of the same subformula within A may receive different assignments.

Definition 3.11 (Potential pre-realizations). The mapping

$$|\cdot| \colon \{o(B) \mid o(B) \text{ is a subformula occurrence within } A\} \to \mathcal{P}(\mathcal{L}_\mathsf{J})$$

from subformula occurrences within our fixed modal formula A to sets of $\mathcal{L}_\mathsf{J}$-formulas is inductively defined as follows:

1. $|o^+(P)| = |o^-(P)| := \{P\}$
   for any occurrence of $P \in \mathsf{Prop}$ (within A);

2. $|o^+(\bot)| = |o^-(\bot)| := \{\bot\}$ for any occurrence of $\bot$;

3. $|o_i^+(B \to C)| := \{B' \to C' \mid B' \in |o_i^-(B)| \text{ and } C' \in |o_i^+(C)|\}$;

4. $|o_i^-(B \to C)| := \{B' \to C' \mid B' \in |o_i^+(B)| \text{ and } C' \in |o_i^-(C)|\}$;

5. $|o_i^+(\Box B)| := \{t{:}(B'_1 \vee \cdots \vee B'_n) \mid t \in \mathsf{Tm} \text{ and } B'_1, \ldots, B'_n \in |o_i^+(B)|\}$;

6. $|o_i^-(\Box B)| := \{x{:}B' \mid x = \mathcal{A}(o_i^-(\Box B)) \text{ and } B' \in |o_i^-(B)|\}$.

3.2 Semantic Realization

The canonical model $\mathcal{M}^c$ can also be viewed as a Kripke structure if one ignores the evaluation function for terms. Even though it is always clear from the context whether $\Vdash$ is used with respect to modular models or Kripke semantics (i.e., with respect to $\mathcal{L}_\mathsf{J}$-formulas or to $\mathcal{L}_\square$-formulas), we will sometimes write $\Vdash_{\mathsf{LP_{CS}}}$ and $\Vdash_{\mathsf{S4}}$ respectively, to emphasize this point.

For a set Γ of formulas of either type, we write

$$\mathcal{M}^c, \Phi \Vdash \Gamma \quad \text{to mean} \quad \mathcal{M}^c, \Phi \Vdash B \text{ for all } B \in \Gamma$$

and

$$\neg \Gamma \quad \text{to mean} \quad \{\neg D \mid D \in \Gamma\} \ .$$

Lemma 3.12. *For any axiomatically appropriate constant specification* CS,

1. *for any* $\Phi \in W^c$,
   $\mathcal{M}^c, \Phi \Vdash_{\mathsf{S4}} \neg B$ *whenever* $\mathcal{M}^c, \Phi \Vdash_{\mathsf{LP_{CS}}} \neg |o^+(B)|$;[2]

2. *for any* $\Phi \in W^c$,
   $\mathcal{M}^c, \Phi \Vdash_{\mathsf{S4}} B$ *whenever* $\mathcal{M}^c, \Phi \Vdash_{\mathsf{LP_{CS}}} |o^-(B)|$.

Proof. The two claims are shown simultaneously by induction on the structure of B for all subformulas B occurring within A.

- If $B = P$ for some $P \in \mathsf{Prop}$, both statements are trivial since $|o^-(P)| = \{P\}$ and $\neg|o^+(P)| = \{\neg P\}$.

- If $B = \bot$ occurs positively within A, the first statement is trivial since $\mathcal{M}^c, \Phi \Vdash_{\mathsf{S4}} \neg \bot$ for each Φ.

- If $B = \bot$ occurs negatively within A, the second statement is immediate because $|o^-(\bot)| = \{\bot\}$ and, hence, $\mathcal{M}^c, \Phi \Vdash_{\mathsf{LP_{CS}}} |o^-(\bot)|$ cannot hold for any Φ.

- For a positive occurrence of $B = D \to C$, assume that

  $$\mathcal{M}^c, \Phi \Vdash_{\mathsf{LP_{CS}}} \neg |o_i^+(D \to C)| \ ,$$

  i.e., that for each $D' \in |o_i^-(D)|$ and each $C' \in |o_i^+(C)|$

  $$\mathcal{M}^c, \Phi \Vdash_{\mathsf{LP_{CS}}} \neg(D' \to C') \ .$$

---

[2] We write $\neg|o^+(B)|$ rather than $|o^-(\neg B)|$ because we need to apply this lemma to $\neg A$, which is not a subformula within A.

Thus, $\mathcal{M}^c, \Phi \Vdash_{\mathsf{LP_{CS}}} D'$ for each $D' \in |o_i^-(D)|$ and $\mathcal{M}^c, \Phi \Vdash_{\mathsf{LP_{CS}}} \neg C'$ for each $C' \in |o_i^+(C)|$. In other words,

$$\mathcal{M}^c, \Phi \Vdash_{\mathsf{LP_{CS}}} |o_i^-(D)| \quad \text{and} \quad \mathcal{M}^c, \Phi \Vdash_{\mathsf{LP_{CS}}} \neg|o_i^+(C)| \ .$$

By the induction hypothesis,

$$\mathcal{M}^c, \Phi \Vdash_{\mathsf{S4}} D \quad \text{and} \quad \mathcal{M}^c, \Phi \Vdash_{\mathsf{S4}} \neg C \ .$$

Hence,
$$\mathcal{M}^c, \Phi \Vdash_{\mathsf{S4}} \neg(D \to C) \ .$$

- For a negative occurrence of $B = D \to C$, assume that

$$\mathcal{M}^c, \Phi \Vdash_{\mathsf{LP_{CS}}} |o_i^-(D \to C)| \ . \tag{3.3}$$

We distinguish two subcases.

  - If $\mathcal{M}^c, \Phi \Vdash_{\mathsf{LP_{CS}}} \neg|o_i^+(D)|$, then $\mathcal{M}^c, \Phi \Vdash_{\mathsf{S4}} \neg D$ by the induction hypothesis. Thus,

  $$\mathcal{M}^c, \Phi \Vdash_{\mathsf{S4}} D \to C \ .$$

  - Otherwise, there is a $D' \in |o_i^+(D)|$ such that $\mathcal{M}^c, \Phi \Vdash_{\mathsf{LP_{CS}}} D'$. For any $C' \in |o_i^-(C)|$, we have $D' \to C' \in |o_i^-(D \to C)|$ and, by our assumption (3.3), $\mathcal{M}^c, \Phi \Vdash_{\mathsf{LP_{CS}}} D' \to C'$. It follows that $\mathcal{M}^c, \Phi \Vdash_{\mathsf{LP_{CS}}} C'$ for each $C' \in |o_i^-(C)|$, i.e.,

  $$\mathcal{M}^c, \Phi \Vdash_{\mathsf{LP_{CS}}} |o_i^-(C)| \ .$$

  By the induction hypothesis, $\mathcal{M}^c, \Phi \Vdash_{\mathsf{S4}} C$, again implying

  $$\mathcal{M}^c, \Phi \Vdash_{\mathsf{S4}} D \to C \ .$$

- For a positive occurrence of $B = \Box C$, assume

$$\mathcal{M}^c, \Phi \Vdash_{\mathsf{LP_{CS}}} \neg|o_i^+(\Box C)| \ .$$

It is sufficient to show that

$$\Phi/\sharp \ \cup \ \neg|o_i^+(C)| \quad \text{is } \mathsf{LP_{CS}}\text{-consistent.} \tag{3.4}$$

Indeed, (3.4) implies that there exists a maximal $\mathsf{LP_{CS}}$-consistent set $\Psi \in W^c$ such that

$$\Phi/\sharp \ \cup \ \neg|o_i^+(C)| \quad \subseteq \quad \Psi \ .$$

3.2 Semantic Realization

By Truth Lemma 3.7 for $\mathcal{M}^c$, given $\neg|o_i^+(C)| \subseteq \Psi$, it follows that $\mathcal{M}^c, \Psi \Vdash_{\mathsf{LP_{CS}}} \neg|o_i^+(C)|$. By the induction hypothesis (at Ψ), we get $\mathcal{M}^c, \Psi \Vdash_{\mathsf{S4}} \neg C$. It remains to note that $R^c(\Phi, \Psi)$ because $\Phi/\sharp \subseteq \Psi$. Hence, $\mathcal{M}^c, \Phi \Vdash_{\mathsf{S4}} \neg\Box C$.

Let us now show the open claim (3.4). Suppose towards a contradiction that $\Phi/\sharp \cup \neg|o_i^+(C)|$ is not $\mathsf{LP_{CS}}$-consistent, i.e., there exist finite sets

$$\Sigma \subseteq \Phi/\sharp \quad \text{and} \quad \{C_1', \ldots, C_n'\} \subseteq |o_i^+(C)|$$

for some $n \geq 0$ such that

$$\mathsf{LP_{CS}} \vdash \bigwedge\bigl(\Sigma \cup \{\neg C_1', \ldots, \neg C_n'\}\bigr) \to \bot \ .$$

Since the set $|o_i^+(C)|$ cannot be empty, we can assume, without loss of generality, that $n \geq 1$. By propositional reasoning,

$$\mathsf{LP_{CS}} \vdash \bigwedge\bigl(\Sigma \cup \{\neg(C_1' \vee \cdots \vee C_n')\}\bigr) \to \bot \ ,$$

so we conclude that

$$\Phi/\sharp \ \cup \ \{\neg(C_1' \vee \cdots \vee C_n')\} \quad \text{is not } \mathsf{LP_{CS}}\text{-consistent.}$$

By Lemma 3.8, given the axiomatic appropriateness of CS, this implies that there is a term t such that

$$t{:}(C_1' \vee \cdots \vee C_n') \in \Phi \ .$$

Hence, by Truth Lemma 3.7 for $\mathcal{M}^c$,

$$\mathcal{M}^c, \Phi \Vdash_{\mathsf{LP_{CS}}} t{:}(C_1' \vee \cdots \vee C_n') \ ,$$

which contradicts the assumption $\mathcal{M}^c, \Phi \Vdash_{\mathsf{LP_{CS}}} \neg|o_i^+(\Box C)|$.

- For a negative occurrence of $B = \Box C$, assume

$$\mathcal{M}^c, \Phi \Vdash_{\mathsf{LP_{CS}}} |o_i^-(\Box C)| \ .$$

Let $\mathcal{A}(o_i^-(\Box C)) = x$. By the assumption, for any $C' \in |o_i^-(C)|$, we have $\mathcal{M}^c, \Phi \Vdash_{\mathsf{LP_{CS}}} x{:}C'$, and, hence, by JYB, $\mathcal{M}^c, \Psi \Vdash_{\mathsf{LP_{CS}}} C'$ for any $\Psi \in W^c$ with $R^c(\Phi, \Psi)$. In other words, $\mathcal{M}^c, \Psi \Vdash_{\mathsf{LP_{CS}}} |o_i^-(C)|$ whenever $R^c(\Phi, \Psi)$. By the induction hypothesis (at Ψ), we obtain $\mathcal{M}^c, \Psi \Vdash_{\mathsf{S4}} C$ whenever $R^c(\Phi, \Psi)$. Finally, this yields

$$\mathcal{M}^c, \Phi \Vdash_{\mathsf{S4}} \Box C \ . \qquad \Box$$

Lemma 3.13. *Let CS be an axiomatically appropriate constant specification. For our fixed formula A, if* $\mathsf{S4} \vdash A$, *then there are*

$$A_1, \ldots, A_n \in |o^+(A)|$$

such that

$$\mathsf{LP_{CS}} \vdash A_1 \vee \cdots \vee A_n \ .$$

Proof. We show the contraposition. Assume that for any

$$A_1, \ldots, A_n \in |o^+(A)| \ ,$$

we have

$$\mathsf{LP_{CS}} \nvdash A_1 \vee \cdots \vee A_n \ ,$$

or, equivalently, using propositional reasoning,

$$\mathsf{LP_{CS}} \nvdash \bigwedge \{\neg A_1, \ldots, \neg A_n\} \to \bot \ .$$

Thus, $\neg|o^+(A)|$ is $\mathsf{LP_{CS}}$-consistent and the canonical model $\mathcal{M}^c$ contains a world $\Phi \in W^c$ such that $\neg|o^+(A)| \subseteq \Phi$. By Truth Lemma 3.7 we get

$$\mathcal{M}^c, \Phi \Vdash_{\mathsf{LP_{CS}}} \neg|o^+(A)| \ .$$

By Lemma 3.12, given the axiomatic appropriateness of CS,

$$\mathcal{M}^c, \Phi \Vdash_{\mathsf{S4}} \neg A \ ,$$

or, equivalently, $\mathcal{M}^c, \Phi \nVdash_{\mathsf{S4}} A$. Since $\mathcal{M}^c$ is not only a model of $\mathsf{LP_{CS}}$ but also of S4, we can use soundness of S4 to conclude $\mathsf{S4} \nvdash A$. $\square$

The next step is to look at *potential realizations* and it remains to be shown that the previously defined potential pre-realizations can be turned into such potential realizations.

Definition 3.14 (Potential realizations)**.** The mapping

$$\|\cdot\| : \{o(B) \mid o(B) \text{ is a subformula occurrence within } A\} \to \mathcal{P}(\mathcal{L}_J)$$

is defined like $|\cdot|$ in Definition 3.11 with $\|\cdot\|$ substituted for $|\cdot|$ everywhere, with the exception of the case of $o^+(\Box B)$, which for $\|\cdot\|$ is defined by

$$\|o_i^+(\Box B)\| := \{t{:}B' \mid B' \in \|o_i^+(B)\| \text{ and } t \in \mathsf{Tm}\} \ .$$

3.2 Semantic Realization

Remark 3.15. It is easy to show by induction on the complexity of subformula occurrences within A that
$$B' \in \|o(B)\| \quad \text{implies} \quad (B')^\circ = B \ .$$
In particular, for each element A' of $\|o^+(A)\|$, the forgetful projection of A' is our fixed formula A. This is the reason we call formulas from $\|o^+(A)\|$ *potential realizations* of A.

Definition 3.16 (Input variable). Let $o_i(B)$ be a (positive or negative) subformula occurrence within our fixed formula A. A justification variable x is called an *input variable* in $o_i(B)$ if $x = \mathcal{A}(o_j^-(\Box C))$ for some suboccurrence of $\Box C$ within $o_i(B)$. Note that, the notation $o_j^-(\Box C)$ means that this occurrence of $\Box C$ is negative within A (but not necessarily negative within $o_i(B)$).

In this chapter, by a *substitution* we mean a *variable substitution*, i.e., a substitution that maps variables to terms but is the identity function on all atomic propositions.

Definition 3.17 (Properties of substitutions). The *domain* $\mathsf{dom}(\sigma)$ of a substitution σ is defined as
$$\mathsf{dom}(\sigma) := \{x \in \mathsf{JVar} \mid x \neq \sigma(x)\} \ .$$
For two substitutions σ_1 and σ_2 with $\mathsf{dom}(\sigma_1) \cap \mathsf{dom}(\sigma_2) = \emptyset$ we define the substitution $\sigma_1 \cup \sigma_2$ by
$$(\sigma_1 \cup \sigma_2)(x) := \begin{cases} \sigma_1(x) & \text{if } x \in \mathsf{dom}(\sigma_1), \\ \sigma_2(x) & \text{if } x \in \mathsf{dom}(\sigma_2), \\ x & \text{otherwise.} \end{cases}$$
For a (positive or negative) subformula occurrence $o(C)$ within our fixed formula A: a substitution σ is said to *live on* $o(C)$ if
$$\mathsf{dom}(\sigma) \subseteq \{x \mid x \text{ is an input variable in } o(C)\} \ ;$$
a substitution σ is said to *live away from* $o(C)$ if
$$\mathsf{dom}(\sigma) \cap \{x \mid x \text{ is an input variable in } o(C)\} = \emptyset \ .$$
A substitution σ satisfies the *no new variable condition* if for any variable x
$$\sigma(x) \text{ contains no variables other than } x.$$

The proofs of the following two lemmas are left to the readers as an exercise.

Lemma 3.18. *Assume we are given two substitutions σ_1 and σ_2 such that for some subformula occurrence $o(C)$ within our fixed formula A*

1. *σ_1 lives on $o(C)$,*

2. *σ_2 lives away from $o(C)$, and*

3. *both σ_1 and σ_2 meet the no new variable condition.*

Then $\sigma_1 \cup \sigma_2$ is defined, is equal to $\sigma_1 \circ \sigma_2 = \sigma_2 \circ \sigma_1$, and meets the no new variable condition.

Lemma 3.19. *Assume $C' \in \|o(C)\|$ and a substitution σ lives away from $o(C)$. Then $C'\sigma \in \|o(C)\|$.*

Remark 3.20. However, the statement that σ lives away from $o(C)$ does not imply that $C' = C'\sigma$ for formulas $C' \in \|o(C)\|$ since variables in the domain of σ still might occur in C'. For instance, $A = \Box P$ has no negatively occurring subformulas of the form $\Box C$, accordingly there are no input variables in the occurrence of $\Box P$ within itself, and every substitution lives away from this occurrence. However, for

$$x{:}P \in \|o^+(\Box P)\|$$

it is trivial to construct a substitution that would change $x{:}P$.

This observation is important in the proof of Lemma 3.21 in the cases for implications and negatively occurring modalities.

Lemma 3.21. *Let CS be an axiomatically appropriate and schematic constant specification. For every subformula occurrence $o_i(B)$ within our fixed formula A and for all*

$$B_1, \ldots, B_n \in |o_i(B)| \ ,$$

there is a formula $B' \in \|o_i(B)\|$ and a substitution σ that lives on $o_i(B)$ and meets the no new variable condition such that

1. *if $o_i(B) = o_i^+(B)$ is a positive occurrence, then*

$$\mathsf{LP}_{\mathsf{CS}} \vdash (B_1 \vee \cdots \vee B_n)\sigma \to B' \ ;$$

3.2 Semantic Realization

2. if $o_i(B) = o_i^-(B)$ is a negative occurrence, then
$$\mathsf{LP}_{\mathsf{CS}} \vdash B' \to (B_1 \wedge \cdots \wedge B_n)\sigma \ .$$

Proof. The two claims are shown simultaneously by induction on the complexity of $o_i(B)$. Throughout the proof, we use the fact that σ distributes through all binary Boolean connectives and commutes with negation.

1. If $B = P$ where $P \in \mathsf{Prop}$, then
$$|o_i^+(P)| = \|o_i^+(P)\| = |o_i^-(P)| = \|o_i^-(P)\| = \{P\}$$

   and, thus, in both statements we can use the empty substitution (i.e., the identity function), using the fact that both
$$P \vee \cdots \vee P \to P \qquad \text{and} \qquad P \to P \wedge \cdots \wedge P$$
   are propositional tautologies.

2. If $B = \bot$, the situation is analagous since
$$|o^+(\bot)| = \|o^+(\bot)\| = |o^-(\bot)| = \|o^-(\bot)\| = \{\bot\} \ .$$
   Thus we can again use the empty substitution.

3. For a positive occurrence of $B = D \to C$, let
$$D_1 \to C_1, \ldots, D_n \to C_n \in |o_i^+(D \to C)| \ .$$
   Then
$$C_1, \ldots, C_n \in |o_i^+(C)| \quad \text{and} \quad D_1, \ldots, D_n \in |o_i^-(D)| \ .$$
   By the induction hypothesis, there are
$$C' \in \|o_i^+(C)\| \ , \quad D' \in \|o_i^-(D)\| \ ,$$
   and substitutions σ_C and σ_D such that
$$\mathsf{LP}_{\mathsf{CS}} \vdash (C_1 \vee \cdots \vee C_n)\sigma_C \to C'$$
   and
$$\mathsf{LP}_{\mathsf{CS}} \vdash D' \to (D_1 \wedge \cdots \wedge D_n)\sigma_D \ .$$

and, moreover, such that both σ_C and σ_D meet the no new variable condition, σ_C lives on $o_i^+(C)$, and σ_D lives on $o_i^-(D)$. Since CS is schematic, we find by Corollary 2.29

$$\mathsf{LP}_{\mathsf{CS}} \vdash \Big((C_1 \vee \cdots \vee C_n)\sigma_C \to C'\Big)\sigma_D \tag{3.5}$$

and

$$\mathsf{LP}_{\mathsf{CS}} \vdash \Big(D' \to (D_1 \wedge \cdots \wedge D_n)\sigma_D\Big)\sigma_C . \tag{3.6}$$

Because substitutions σ_C and σ_D live on disjoint subformula occurrences within A and because annotation $\mathcal{A}$ is an injective mapping, σ_C lives away from $o_i^-(D)$ (and also σ_D lives away from $o_i^+(C)$). By Lemma 3.18, we can define

$$\sigma := \sigma_C \cup \sigma_D = \sigma_C \circ \sigma_D = \sigma_D \circ \sigma_C ,$$

which meets the no new variable condition. Thus, (3.5) and (3.6) can be rewritten as

$$\mathsf{LP}_{\mathsf{CS}} \vdash (C_1 \vee \cdots \vee C_n)\sigma \to C'\sigma_D$$

and

$$\mathsf{LP}_{\mathsf{CS}} \vdash D'\sigma_C \to (D_1 \wedge \cdots \wedge D_n)\sigma .$$

Using propositional reasoning, we obtain

$$\mathsf{LP}_{\mathsf{CS}} \vdash \Big((D_1 \to C_1) \vee \cdots \vee (D_n \to C_n)\Big)\sigma \to (D'\sigma_C \to C'\sigma_D) .$$

By Lemma 3.19, we have $D'\sigma_C \in \|o_i^-(D)\|$ and $C'\sigma_D \in \|o_i^+(C)\|$, meaning that

$$D'\sigma_C \to C'\sigma_D \in \|o_i^+(D \to C)\| .$$

It remains to prove that σ lives on $o_i^+(D \to C)$, which is rather simple and left to the reader. Thus, we are done with this case.

4. $B = D \to C$ occurring negatively. This case is symmetric to the previous case and thus omitted.

5. For a positive occurrence of $B = \Box C$, let

$$t_1{:}C_1, \ldots, t_n{:}C_n \in |o_k^+(\Box C)| .$$

3.2 Semantic Realization

By the definition of $|\cdot|$, for each $1 \leq i \leq n$,

$$C_i = C_{i,1} \vee C_{i,2} \vee \cdots \vee C_{i,m_i} ,$$

where $m_i \geq 1$ and $C_{i,j} \in |o_k^+(C)|$ for each $1 \leq j \leq m_i$. By the induction hypothesis for the collection of all these $m_1 + \cdots + m_n$ formulas, there exists a formula $C' \in \|o_k^+(C)\|$ and a substitution σ that lives on $o_k^+(C)$ and meets the no new variable condition such that

$$\mathsf{LP_{CS}} \vdash (C_{1,1} \vee \cdots C_{1,m_1} \vee \cdots \vee C_{n,1} \vee \cdots \vee C_{n,m_n})\sigma \to C' .$$

Using propositional reasoning, we can group disjuncts to obtain

$$\mathsf{LP_{CS}} \vdash (C_1 \vee \cdots \vee C_n)\sigma \to C' ,$$

and, hence, for each $1 \leq i \leq n$,

$$\mathsf{LP_{CS}} \vdash C_i\sigma \to C' .$$

Given the axiomatic appropriateness of CS, by Constructive Necessitation there are ground terms r_i such that for each $1 \leq i \leq n$

$$\mathsf{LP_{CS}} \vdash r_i{:}(C_i\sigma \to C') .$$

For each $1 \leq i \leq n$, using the instance

$$r_i{:}(C_i\sigma \to C') \to \left(t_i\sigma{:}C_i\sigma \to r_i \cdot (t_i\sigma){:}C'\right)$$

of the axiom scheme **j**, we find by (**MP**) that

$$\mathsf{LP_{CS}} \vdash t_i\sigma{:}C_i\sigma \to r_i \cdot (t_i\sigma){:}C' .$$

We set $s := r_1 \cdot (t_1\sigma) + \cdots + r_n \cdot (t_n\sigma)$ and, using several instances of the axiom scheme **j+** and equalities $(t_i{:}C_i)\sigma = t_i\sigma{:}C_i\sigma$, we obtain for each $1 \leq i \leq n$ that

$$\mathsf{LP_{CS}} \vdash (t_i{:}C_i)\sigma \to s{:}C' .$$

By propositional reasoning,

$$\mathsf{LP_{CS}} \vdash (t_1{:}C_1)\sigma \vee \cdots \vee (t_n{:}C_n)\sigma \to s{:}C' ,$$

in other words,
$$\mathsf{LP_{CS}} \vdash (t_1{:}C_1 \vee \cdots \vee t_n{:}C_n)\sigma \to s{:}C' \ .$$

Obviously, $s{:}C' \in \|o_k^+(\Box C)\|$. Since the substitution σ meets the no new variable condition, it remains to note that σ living on $o_k^+(\Box C)$ is equivalent to σ living on $o_k^+(C)$ because the outer $\Box$ of $o_k^+(\Box C)$ does not create any new input variables.

6. For a negative occurrence of $B = \Box C$, let
$$x{:}C_1, \ldots, x{:}C_n \in |o_k^-(\Box C)| \ ,$$

where $x = \mathcal{A}(o_k^-(\Box C))$ and $C_1, \ldots, C_n \in o_k^-(C)$. By the induction hypothesis, there is a $C' \in \|o_k^-(C)\|$ and a substitution σ' that meets the no new variable condition and lives on $o_k^-(C)$ such that
$$\mathsf{LP_{CS}} \vdash C' \to (C_1 \wedge \cdots \wedge C_n)\sigma' \ .$$

For each $1 \leq i \leq n$, by propositional reasoning,
$$\mathsf{LP_{CS}} \vdash C' \to C_i\sigma' \ .$$

We define a substitution
$$\sigma_x(y) := \begin{cases} s \cdot x & \text{if } y = x, \\ y & \text{if } y \neq x. \end{cases}$$

Since CS is schematic, we find by Corollary 2.29 that for each $1 \leq i \leq n$
$$\mathsf{LP_{CS}} \vdash (C' \to C_i\sigma')\sigma_x \ .$$

Defining $\sigma := \sigma_x \circ \sigma'$, we can rewrite it for each $1 \leq i \leq n$ as
$$\mathsf{LP_{CS}} \vdash C'\sigma_x \to C_i\sigma \ .$$

Thus, by Constructive Necessitation, for our axiomatically appropriate CS, there are ground terms t_i such that for each $1 \leq i \leq n$
$$\mathsf{LP_{CS}} \vdash t_i{:}(C'\sigma_x \to C_i\sigma) \ .$$

We set $s := t_1 + \cdots + t_n$ and, using propositional reasoning and several instances of the axiom scheme **j+**, we obtain for each $1 \leq i \leq n$
$$\mathsf{LP_{CS}} \vdash s{:}(C'\sigma_x \to C_i\sigma) \ .$$

3.2 Semantic Realization

Using the axiom scheme **j** and (**MP**), we find for each $1 \leq i \leq n$

$$\mathsf{LP}_{\mathsf{CS}} \vdash x{:}(C'\sigma_x) \to s \cdot x{:}(C_i\sigma) \ . \tag{3.7}$$

Given that annotation $\mathcal{A}$ is injective and that $o_k^-(\Box C)$ does not occur within $o_k^-(C)$, it follows that x is not an input variable in $o_k^-(C)$. Since σ' lives on $o_k^-(C)$, we conclude that $x \notin \mathrm{dom}(\sigma')$. Consequently,

$$x\sigma = (x\sigma')\sigma_x = x\sigma_x = s \cdot x \ .$$

For each $1 \leq i \leq n$

$$s \cdot x{:}(C_i\sigma) = x\sigma{:}(C_i\sigma) = (x{:}C_i)\sigma \ ,$$

and (3.7) can be rewritten as

$$\mathsf{LP}_{\mathsf{CS}} \vdash x{:}(C'\sigma_x) \to (x{:}C_i)\sigma \ .$$

Hence, by propositional reasoning,

$$\mathsf{LP}_{\mathsf{CS}} \vdash x{:}(C'\sigma_x) \to (x{:}C_1 \wedge \cdots \wedge x{:}C_n)\sigma \ .$$

Clearly, $\mathrm{dom}(\sigma_x) = \{x\}$. Since x is not an input variable in $o_k^-(C)$, we conclude that σ_x lives away from $o_k^-(C)$. Thus, by Lemma 3.19, we have $C'\sigma_x \in \|o_k^-(C)\|$. It remains to show that σ meets the no new variable condition and lives on $o_k^-(\Box C)$. The former statement follows from Lemma 3.18 because both σ' (by the induction hypothesis) and σ_x (obviously) meet the no new variable condition and, as we discussed, σ' lives on $o_k^-(C)$ while σ_x lives away from $o_k^-(C)$. The latter statement follows from the fact that $\sigma(y) = y\sigma'\sigma_x \neq y$ implies that either $\sigma'(y) \neq y$ or $\sigma_x(y) \neq y$, i.e., either $y \in \mathrm{dom}(\sigma')$ or $y = x$, implying in turn that y is either an input variable in $o_k^-(C)$ or x, which is a complete description of all input variables in $o_k^-(\Box C)$. □

Theorem 3.22 (Semantic Realization)**.** *Let* CS *be an axiomatically appropriate and schematic constant specification. If* S4 $\vdash A$*, then there exists an* $\mathcal{L}_\mathsf{J}$*-formula* A' *such that* $(A')^\circ = A$ *and* $\mathsf{LP}_{\mathsf{CS}} \vdash A'$.

Proof. Assume S4 $\vdash A$. In view of Remark 3.15 it is enough to show that there exists $A' \in \|o^+(A)\|$ such that $\mathsf{LP}_{\mathsf{CS}} \vdash A'$. Given that

CS is axiomatically appropriate, by Lemma 3.13 there are formulas $A_1, \ldots, A_n \in |o^+(A)|$ such that

$$\mathsf{LP}_{\mathsf{CS}} \vdash A_1 \vee \cdots \vee A_n \; . \tag{3.8}$$

Given that CS is additionally schematic, by Lemma 3.21 there is a formula $A' \in \|o^+(A)\|$ and a substitution σ such that

$$\mathsf{LP}_{\mathsf{CS}} \vdash (A_1 \vee \cdots \vee A_n)\sigma \to A' \; . \tag{3.9}$$

Since CS is schematic, we obtain from (3.8) and Corollary 2.29 that

$$\mathsf{LP}_{\mathsf{CS}} \vdash (A_1 \vee \cdots \vee A_n)\sigma \; ,$$

which together with (3.9) yields $\mathsf{LP}_{\mathsf{CS}} \vdash A'$. □

3.3 Constructive Realization

The proof of Theorem 3.22 relies on the fact that a modular model can also be viewed as a Kripke structure. Thus we can evaluate both justification logic formulas and modal logic formulas within the same model, which makes it possible to establish the Realization Theorem. This semantic proof, however, only shows the existence of a realization. It does not provide a construction of a realization $r(F)$ for a given S4-theorem F.

In this section, we show an alternative proof of the Realization Theorem that uses proof-theoretic means. This proof is constructive and provides an algorithm for building the wanted realization $r(F)$.

First, we need a generalization of the internalization property stated in Lemma 2.18, which is called *Lifting Lemma*.

Lemma 3.23 (Lifting Lemma). *Let CS be an axiomatically appropriate constant specification. For arbitrary formulas*

$$A, B_1, \ldots, B_n, s_1{:}C_1, \ldots, s_m{:}C_m$$

of $\mathcal{L}_\mathsf{J}$, if

$$B_1, \ldots, B_n, s_1{:}C_1, \ldots, s_m{:}C_m \vdash_{\mathsf{LP}_{\mathsf{CS}}} A \; ,$$

then there is a term $t(y_1, \ldots, y_n, z_1, \ldots, z_m) \in \mathsf{Tm}$ such that

$$y_1{:}B_1, \ldots, y_n{:}B_n, s_1{:}C_1, \ldots, s_m{:}C_m \vdash_{\mathsf{LP}_{\mathsf{CS}}} t(y_1, \ldots, y_n, s_1, \ldots, s_m){:}A$$

for arbitrary variables $y_1, \ldots, y_n$.

3.3 Constructive Realization

Proof. The proof of the Lifting Lemma is essentially the same proof as the proof of Lemma 2.18 but with one addition case in case distinction:

5. If $A = s_i{:}C_i$, we set $t := !s_i$. Then

$$y_1{:}B_1, \ldots, y_n{:}B_n, s_1{:}C_1, \ldots, s_m{:}C_m \vdash_{\mathsf{LP_{CS}}} !s_i{:}s_i{:}C_i$$

follows using the **j4** axiom. □

In order to establish the constructive Realization Theorem, we need a Gentzen-style deductive system for **S4**. In the following we introduce *multisets* and *sequents* as well as the *system* **GS4**.

Definition 3.24. A *multiset* M over a set A is a mapping $M : A \to \mathbb{N}$.

We may consider a multiset M over A as a collection of elements of A that can contain multiple instances of each $a \in A$. The natural number $M(a)$ states how many instances of a the multiset M contains. Thus, informally, a multiset is a set where the number of instances of its elements matters (but not the order of its elements).

Let M and N be multisets over a set A. By $M \uplus N$ we denote the multiset union of M and N. For each $a \in A$ we set

$$(M \uplus N)(a) := M(a) + N(a) \ .$$

If the set underlying set A is clear from the context, we can employ the usual set notation also for multisets and write, for instance, $\{a, a, b\}$ for a multiset M with $M(a) = 2$ and $M(b) = 1$.

In this section, we mainly consider *finite multisets of formulas*, which will be denoted by Γ, Δ, Σ, and Π. It will be convenient to write Γ, Δ instead of $\Gamma \uplus \Delta$. We will also identify a formula F with the singleton multiset $\{F\}$ and write Γ, F for $\Gamma \uplus \{F\}$.

Definition 3.25. A *sequent* is a expression $\Gamma \supset \Delta$ where Γ and Δ are multisets of formulas.

We may informally read a sequent

$$A_1, \ldots, A_m \supset B_1, \ldots, B_n$$

as the formula

$$(A_1 \wedge \cdots \wedge A_m) \to (B_1 \vee \cdots \vee B_n) \ .$$

$$P \supset P \qquad \bot \supset$$

$$\dfrac{\Gamma \supset \Delta, A \quad B, \Gamma \supset \Delta}{A \to B, \Gamma \supset \Delta} \ (\to \supset) \qquad \dfrac{A, \Gamma \supset \Delta, B}{\Gamma \supset \Delta, A \to B} \ (\supset \to)$$

$$\dfrac{A, \Gamma \supset \Delta}{\Box A, \Gamma \supset \Delta} \ (\Box \supset) \qquad \dfrac{\Box \Gamma \supset A}{\Box \Gamma \supset \Box A} \ (\supset \Box)$$

$$\dfrac{\Gamma \supset \Delta}{A, \Gamma \supset \Delta} \ (\text{w} \supset) \qquad \dfrac{\Gamma \supset \Delta}{\Gamma \supset \Delta, A} \ (\supset \text{w})$$

$$\dfrac{A, A, \Gamma \supset \Delta}{A, \Gamma \supset \Delta} \ (\text{c} \supset) \qquad \dfrac{\Gamma \supset \Delta, A, A}{\Gamma \supset \Delta, A} \ (\supset \text{c})$$

Figure 3.1: Axioms and rules of the sequent system GS4

The sequent system GS4 [112] derives sequents of $\mathcal{L}_\Box$ formulas. It consists of the axioms and rules that are shown in Figure 3.1. Note that in our formulation of GS4, we use axioms of the form $P \supset P$ whereas in [112] axioms $A \supset A$ are employed. This is fine since in our formulation $A \supset A$ is derivable for any formula A. The reason for this change is that the proof of Theoren 3.31 requires axioms to consist of atomic propositions only. We write $\vdash_{\mathsf{GS4}} \Gamma \supset \Delta$ if the sequent $\Gamma \supset \Delta$ is provable in GS4. It is standard to show that GS4 is sound and complete [112, 136].

Theorem 3.26. *Let A be a formula of $\mathcal{L}_\Box$. We have that*

$$\vdash_{\mathsf{GS4}} \supset A \quad \text{if and only if} \quad A \text{ is valid}.$$

For the constructive realization proof, we will distinguish positive and negative occurrences of $\Box$ in a sequent $\Gamma \supset \Delta$.

Definition 3.27 (Polarity of $\Box$). Let $\Box B$ be a subformula occurrence within a formula $A \in \Delta$. The $\Box$-operator of the subformula occurrence $\Box B$ in the sequent $\Gamma \supset \Delta$ has the same *polarity* as the subformula occurrence of $\Box B$ within A.

Let $\Box B$ be a subformula occurrence within a formula $A \in \Gamma$. The $\Box$-operator of the subformula occurrence $\Box B$ in the sequent $\Gamma \supset \Delta$ has the opposite *polarity* as the subformula occurrence of $\Box B$ within A.

Remark 3.28. Observe that the rules of GS4 respect the polarities of $\Box$-operators so that the rule $(\supset \Box)$ introduces positive occurrences of $\Box$ and the rule $(\Box \supset)$ introduces negative occurrences of $\Box$.

3.3 Constructive Realization

Definition 3.29 (Essential family of □-occurrences). Let $\mathcal{D}$ be a derivation in GS4. We say that occurrences of □ in $\mathcal{D}$ are *related* if they occur at the same position in related formulas of premises and conclusions of a rule instance in $\mathcal{D}$; we close this relationship of related occurrences under transitivity.

All occurrences of □ in $\mathcal{D}$ naturally split into disjoint *families* of related occurrences.

We call a family *essential* if at least one of its members is introduced by a $(\supset \Box)$ rule. Note that an essential family is positive (i.e. contains only positive occurrences).

Definition 3.30 (Normal realization). A realization is *normal* if all negative occurrences of □ are realized by distinct justification variables.

Theorem 3.31 (Constructive realization). *Let CS be an axiomatically appropriate and schematic constant specification. Then there exists a normal realization r such that for each $\mathcal{L}_\mathsf{J}$-formula A we have*

$$\vdash_{\mathsf{GS4}} \supset A \quad \textit{implies} \quad \vdash_{\mathsf{LP_{CS}}} r(A).$$

Proof. Let $\mathcal{D}$ be the GS4 derivation that proves $\supset A$. The desired realization r is constructed by the following three steps. We reserve a large enough set of justification variables as *provisional variables*.

1. For each negative family and each non-essential positive family, replace all □ occurrences by x: where we choose a fresh justification variable for each family.

2. Pick an essential family f. Enumerate all occurrences of $(\supset \Box)$ rules that introduce a □-operator to this family. Let n_f denote the number of such occurrences. Replace each □ with a justification term
$$v_1 + \cdots + v_{n_f}$$
where each v_i is a fresh provisional variable. Do this for each essential family. The resulting tree $\mathcal{D}'$ is labelled by $\mathcal{L}_\mathsf{J}$-formulas.

3. Replace the provisional variables in $\mathcal{D}'$ starting with the leaves and working toward the root. By induction on the depth of a node in $\mathcal{D}'$ we establish that after the process passes a node, the sequent assigned to this node becomes derivable in $\mathsf{LP_{CS}}$ where

derivability of $\Gamma \supset \Delta$ means[3]

$$\Gamma \vdash_{\mathsf{LP_{CS}}} \bigvee \Delta \ .$$

We distinguish the following cases.

(a) The axioms $P \supset P$ and $\bot \supset$ are derivable in $\mathsf{LP_{CS}}$ since we have $P \vdash_{\mathsf{LP_{CS}}} P$ and $\bot \vdash_{\mathsf{LP_{CS}}} \bot$.[4]

(b) For every rule other than $(\supset \Box)$ we do not change the term assignment and establish that the conclusion of the rule is derivable in $\mathsf{LP_{CS}}$ if the premises are. Let us show only the case for $(\to \supset)$. By I.H. we have

$$\Gamma \vdash_{\mathsf{LP_{CS}}} A \vee \bigvee \Delta \ .$$

Hence

$$A \to B, \Gamma \vdash_{\mathsf{LP_{CS}}} A \vee \bigvee \Delta$$

and thus

$$A \to B, \Gamma \vdash_{\mathsf{LP_{CS}}} B \vee \bigvee \Delta \ . \qquad (3.10)$$

By I.H. we also have

$$B, \Gamma \vdash_{\mathsf{LP_{CS}}} \bigvee \Delta \ .$$

By the Deduction Theorem we get

$$\Gamma \vdash_{\mathsf{LP_{CS}}} B \to \bigvee \Delta$$

and thus

$$A \to B, \Gamma \vdash_{\mathsf{LP_{CS}}} B \to \bigvee \Delta \ .$$

Together with (3.10) we finally get

$$A \to B, \Gamma \vdash_{\mathsf{LP_{CS}}} \bigvee \Delta \ .$$

(c) Let an occurrence of a $(\supset \Box)$ rule have number i in the enumeration of all $(\supset \Box)$ rules in a given family f. The corresponding node in $\mathcal{D}'$ is labelled by

$$\frac{y_1{:}B_1, \ldots, y_k{:}B_k \supset A}{y_1{:}B_1, \ldots, y_k{:}B_k \supset u_1 + \cdots + u_{n_f}{:}A}$$

---

[3] Strictly speaking Γ is a multiset. Here we ignore the multiplicities of its elements and consider it as an ordinary set.
[4] Recall that $\bigvee \emptyset := \bot$.

where the y's are justification variables, the u's are justification terms, and u_i is a provisional variable. By the induction hypothesis
$$y_1{:}B_1, \ldots, y_k{:}B_k \supset A$$
is derivable in $\mathsf{LP_{CS}}$. Using Lemma 3.23, we construct a term t such that
$$y_1{:}B_1, \ldots, y_k{:}B_k \vdash_{\mathsf{LP_{CS}}} t{:}A \ .$$

Thus
$$y_1{:}B_1, \ldots, y_k{:}B_k \vdash_{\mathsf{LP_{CS}}} u_1 + \cdots + u_{i-1} + t + u_{i+1} + \cdots + u_{n_f}{:}A \ .$$

Substitute t for u_i everywhere in $\mathcal{D}'$. Since CS is schematic, we find by Corollary 2.29 that this substitution does not affect the already established derivability results.

Eventually, all provisional variables are replaced with terms of non-provisional variables in $\mathcal{D}'$ and we have established that its root sequent $r(A)$ is derivable in $\mathsf{LP_{CS}}$. The realization r built by this construction is normal. □

This is a constructive proof of the Realization Theorem. It does not only establish the existence of a realization but provides an algorithm for constructing it. Moreover, the constructed realization is normal, i.e. negative occurrences of □ are realized by justification variables and positive occurrences of □ are realized by justification terms that depend on these variables. This corresponds to the fact that the negative occurrences can be seen as assumptions from which the positive occurrences are derived.

Example 3.32. In this example we study the realization procedure for the formula
$$\Box A \vee \Box B \to \Box (A \vee B) \ ,$$
see also Example 2.11.

First we recall that $A \vee B$ is an abbreviation for $(A \to \bot) \to B$. Therefore, the rule
$$\frac{\Gamma \supset \Delta, A, B}{\Gamma \supset \Delta, A \vee B}$$
is derivable. Indeed, we have

$$\dfrac{\Gamma\supset\Delta,A,B\quad\dfrac{}{\Gamma,\bot\supset\Delta,B}\bot\supset}{\dfrac{\Gamma,A\to\bot\supset\Delta,B}{\Gamma\supset\Delta,(A\to\bot)\to B}}$$

Simiarly, the rule

$$\dfrac{\Gamma,A\supset\Delta\quad\Gamma,B\supset\Delta}{\Gamma,A\vee B\supset\Delta}$$

is derivable. Indeed, we have

$$\dfrac{\dfrac{\dfrac{\dfrac{\Gamma,A\supset\Delta}{\Gamma,A\supset\Delta,\bot}}{\Gamma\supset\Delta,A\to\bot}\quad\Gamma,B\supset\Delta}{\Gamma,(A\to\bot)\to B\supset\Delta}}{}$$

Using these two derived rules we find the following **GS4**-proof

$$\dfrac{\dfrac{\dfrac{\dfrac{\dfrac{A\supset A}{A\supset A,B}}{\dfrac{A\supset A\vee B}{\Box A\supset A\vee B}}}{\Box A\supset\Box(A\vee B)}(1)\quad\dfrac{\dfrac{\dfrac{\dfrac{B\supset B}{B\supset A,B}}{\dfrac{B\supset A\vee B}{\Box B\supset A\vee B}}}{\Box B\supset\Box(A\vee B)}(2)}{\dfrac{\Box A\vee\Box B\supset\Box(A\vee B)}{\supset(\Box A\vee\Box B)\to\Box(A\vee B)}}}{}$$

The families of $\Box$-occurrences in the formulas $\Box A$ and $\Box B$, respectively, are not essential. The familiy of $\Box$-occurrences in $\Box(A\vee B)$ is essential and there are two instances of $(\supset\Box)$ that introduce a $\Box$-operator in this family. They are marked with (1) and (2), respectively.

After the first two steps of the realization procedure we have the following tree:

$$\dfrac{\dfrac{\dfrac{\dfrac{\dfrac{A\supset A}{A\supset A,B}}{\dfrac{A\supset A\vee B}{x{:}A\supset A\vee B}}}{x{:}A\supset v_1+v_2{:}(A\vee B)}(1)\quad\dfrac{\dfrac{\dfrac{\dfrac{B\supset B}{B\supset A,B}}{\dfrac{B\supset A\vee B}{y{:}B\supset A\vee B}}}{y{:}B\supset v_1+v_2{:}(A\vee B)}(2)}{\dfrac{x{:}A\vee y{:}B\supset v_1+v_2{:}(A\vee B)}{\supset(x{:}A\vee y{:}B)\to v_1+v_2{:}(A\vee B)}}}{}$$

3.3 Constructive Realization

In the last step of the realization procedure, when we deal with rule (1), we find by the induction hypothesis that

$$x{:}A \vdash_{\mathsf{LP_{CS}}} A \vee B \ .$$

By the Lifting Lemma, there is a term $t(x)$ with

$$x{:}A \vdash_{\mathsf{LP_{CS}}} t(x){:}(A \vee B) \ .$$

Substituting $t(x)$ for v_1 yields

$$\dfrac{\dfrac{\dfrac{\dfrac{A \supset A}{A \supset A, B}}{\dfrac{A \supset A \vee B}{x{:}A \supset A \vee B}}}{x{:}A \supset t(x) + v_2{:}(A \vee B)} \ (1) \qquad \dfrac{\dfrac{\dfrac{\dfrac{B \supset B}{B \supset A, B}}{\dfrac{B \supset A \vee B}{y{:}B \supset A \vee B}}}{y{:}B \supset t(x) + v_2{:}(A \vee B)} \ (2)}{\dfrac{x{:}A \vee y{:}B \supset t(x) + v_2{:}(A \vee B)}{\supset (x{:}A \vee y{:}B) \to t(x) + v_2{:}(A \vee B)}}$$

Here it is important that the substitution is done globally on the whole proof. In particular it is also performed in the right branch of the proof although x does not occur in the assumptions of that branch.

Treating rule (2) in the same way yields a term $s(y)$ such that finally

$$\mathsf{LP_{CS}} \vdash (x{:}A \vee y{:}B) \to t(x) + s(y){:}(A \vee B) \ .$$

Example 3.33. The logic S4 features a certain form of self-referentiality. In this example we show how it can occur and how the realization procedure handles it. For more details about this phenomenon we refer to Chapter 7.

We consider a GS4-proof of the formula

$$\Box((P \to \Box P) \to \bot) \to \bot \tag{3.11}$$

that contains two families of $\Box$-occurrences. We highlight them by adding the subscripts 1 and 2, respectively, to the $\Box$-occurrences.

$$\cfrac{\cfrac{\cfrac{\cfrac{\cfrac{\cfrac{\cfrac{\cfrac{\cfrac{P \supset \Box_2 P, P}{\supset P \to \Box_2 P, P} \quad \bot \supset}{(P \to \Box_2 P) \to \bot \supset P}}{\Box_1((P \to \Box_2 P) \to \bot) \supset P}}{\Box_1((P \to \Box_2 P) \to \bot) \supset \Box_2 P}}{\Box_1((P \to \Box_2 P) \to \bot), P \supset \Box_2 P}}{\Box_1((P \to \Box_2 P) \to \bot) \supset P \to \Box_2 P \quad \bot \supset}}{\Box_1((P \to \Box_2 P) \to \bot), (P \to \Box_2 P) \to \bot \supset}}{\Box_1((P \to \Box_2 P) \to \bot), \Box_1((P \to \Box_2 P) \to \bot) \supset}}{\cfrac{\cfrac{\Box_1((P \to \Box_2 P) \to \bot) \supset}{\Box_1((P \to \Box_2 P) \to \bot) \supset \bot}}{\supset \Box_1((P \to \Box_2 P) \to \bot) \to \bot}} \quad (3)$$

Wee see that the family of $\Box_2$-occurrences is essential whereas the family of $\Box_1$-occurrences is not essential.

Consider the rule (3) where $\Box_2$ is introduced. Here we can see the inherent self-referentiality of GS4. Namely, the introduction of $\Box_2$ on the right relies on an assumption (on the left) that includes $\Box_2$. The realization algorithm treats this as follows.

By induction hypothesis we have

$$y{:}((P \to u{:}P) \to \bot) \vdash_{\mathsf{LP}_{\mathsf{CS}}} P$$

where u is a provisional variable. We find a term $t(y)$ such that

$$y{:}((P \to u{:}P) \to \bot) \vdash_{\mathsf{LP}_{\mathsf{CS}}} t(y){:}P \ .$$

Substituting $t(y)$ for u yields

$$y{:}((P \to t(y){:}P) \to \bot) \vdash_{\mathsf{LP}_{\mathsf{CS}}} t(y){:}P \ .$$

3.4 Notes

This proof of Theorem 3.22 mostly follows Fitting's original paper [53], except that the treatment of the cases for implication, both positive and negative, and for negative modality follow [55]. The reason for that is a minor problem discovered by Fitting himself in the above mentioned cases. This is how Fitting describes the problem in [55]:

> "The original proof (but not the result) [of [53, Proposition 7.8]] contains an error, manifesting itself in the Positive and

3.4 Notes

Negative Implication parts, and in the Negative Necessity part. That error has been corrected here, and by ignoring nominals the present proof provides a correct proof of the earlier result."

More specifically, the problem is that in these cases it is necessary to apply a substitution to the potential realization(s) obtained from the induction hypothesis. The original proof from [53] claimed that the substitutions used do not change these potential realizations, whereas the current version postulates instead that the substitutions yield different potential realizations.

Artemov already established a realization theorem in his original paper on the Logic of Proofs [6]. The proof of Theorem 3.31 essentially follows his presentation. Artemov's proof can easily be adapted to realize other modal logics in corresponding justification logics if a cut-free sequent calculus for the modal logic is available, see, e.g., [3, 105].

Brezhnev and Kuznets [32] show that Artemov's algorithm can produce justification terms of exponential length in the size of the initial S4-derivation. They then introduce a modification that produces justification terms of at most quadratic length. Kuznets [95] presents a variant of Artemov's proof of realization that does not need the +-operation on justification terms. The so obtained realization, however, is not normal.

Artemov's procedure to compute realizations is sometimes called *global* realization method since in the case of a ($\supset$ $\Box$) rule, it performs substitutions in the whole derivation. Fitting [54] introduces another proof-theoretic realization method that only performs *local* proof-transformations. Adapting this method to nested sequent systems yields a modular realization theorem that covers many modal logics [31, 74].

In some cases, where there is no cut-free sequent system available for a given modal logic, the following approach may be used to find a realization: translate the modal logic to a logic for which there is a cut-free system available, realize this logic in a suitable justification logic, and apply an inverse translation to obtain a realization in the target justification logic. This approach has been used to realize, e.g., S5 [56] and public announcement logics [36, 38].

Fitting [57] develops a very general realization method that uses the model existence theorem. He applies it to obtain a realization of first-order S4 into a first-order version of the Logic of Proofs.

Chapter 4

Decidability

We study decidability for the Logic of Proofs $\mathsf{LP}_{\mathsf{CS}}$. First we recall that the finite model property does not by itself yield decidability of a logic. Additionally, one needs the satisfaction relation on the finite models to be decidable. As it turns out, this is the main issue in showing decidability for $\mathsf{LP}_{\mathsf{CS}}$. We solve this problem by introducing the class of generated models for $\mathsf{LP}_{\mathsf{CS}}$, in which the interpretation of justification terms is generated by a least fixed point construction. Then, finite generated models, which we call finitary models, provide a decidable satisfaction relation. Showing soundness and completeness of $\mathsf{LP}_{\mathsf{CS}}$ with respect to finitary models yields decidability of $\mathsf{LP}_{\mathsf{CS}}$.

4.1 Post's theorem

In this chapter we study decidability for justification logics. That is the question whether there is an algorithm that given any formula F as input can decide whether F is provable in $\mathsf{LP}_{\mathsf{CS}}$.

For this purpose, we use an informal notion of decidability. A precise formal definition of decidable, i.e. recursive, sets of natural numbers is presented in Chapter 6 on arithmetical interpretations.

For this chapter we simply assume that algorithms take words over some given alphabet as input (and that they produce them as output). Hence the input may consist of justification terms or of formulas. If we assume that the alphabet also includes a delimiter symbol, then the input may consist of pairs of words and even finite sequences of words. For instance the input to an algorithm may be a finite sequence of the

form
$$c_1{:}F_1, c_2{:}F_2, \ldots, c_n{:}F_n \ .$$

We say that

1. *a set S of words is decidable* if there is an algorithm that given any x as input returns *yes* if $x \in S$ and *no* otherwise;

2. *a relation R of words is decidable* if R viewed as a set, i.e. as $\{x \mid R(x)\}$, is decidable;

3. *a set S of words is recursively enumerable* if there is an algorithm that successively outputs all elements of S.

There is an important connection between recursive enumerability and decidability, which is established by Post's theorem.

Theorem 4.1 (Post's Theorem). *If both a set and its complement are recursively enumerable, then the set is decidable.*

For a proof of this theorem, we refer to [109]. The basic idea is the following. Assume we are given a set S and some x. In order to decide whether $x \in S$, we run two programs P_S and $P_{\neg S}$ in parallel where the program P_S enumerates the elements of S and the program $P_{\neg S}$ enumerates the complement of S. In this setting, one of the programs will eventually output x. If x is in the output of P_S, then we return *yes*; if x is in the output of $P_{\neg S}$, then we return *no*. Hence S is decidable.

Definition 4.2. We call a logic L *decidable*, if the set of its theorems, i.e. the set $\{F \mid \mathsf{L} \vdash F\}$, is decidable.

In modal logic, decidability often is established as a consequence of the finite model property [28]. For S4 we can formulate the finite model property as follows: a Kripke model $\mathcal{M} = (W, R, \mathsf{val})$ is called *finite* if

1. W is a finite set and

2. $\mathsf{val}(P) \neq \emptyset$ for only finitely many P.

The finite model property states that S4 is complete with respect to finite Kripke models.[1] This is an important result because finite Kripke models have the following two properties:

---

[1] In many textbooks, the finite model property is stated as completeness with respect to Kripke models with a finite set of possible worlds but no restriction on the valuation function. This, however, is not sufficient for decidability, see, e.g., the discussion in [92].

4.1 Post's theorem

1. The class of finite Kripke models is recursively enumerable.

2. The binary relation $\mathcal{M} \Vdash F$ between finite Kripke models and formulas is decidable.

Hence we obtain decidability of S4 using the following lemma, which is often the key for establishing decidability of a logic.

Lemma 4.3. *Let a finitely axiomatizable logic* L *be sound and complete with respect to a class of models* $\mathbb{C}$, *such that*

1. *the class* $\mathbb{C}$ *is recursively enumerable, and*

2. *the binary relation* $\mathcal{M} \Vdash F$ *between formulas and models from* $\mathbb{C}$ *is decidable.*

Then L *is decidable.*

Proof. Of course, the set of theorems of a finitely axiomatizable logic is recursively enumerable.

Here is an algorithm to recursively enumerate its complement. Since both the set of formulas and the models $\mathcal{M}$ of $\mathbb{C}$ are recursively enumerable, there is an enumeration of all pairs $(\mathcal{M}, F)$. For each pair in this enumeration, the algorithm checks whether $\mathcal{M} \Vdash \neg F$. If it is, the algorithm outputs F, otherwise it skips to the next pair. In this way, the algorithm will enumerate all non-theorems of L. Hence the complement of L is recursively enumerable, too.

Thus by Post's Theorem, L is decidable. □

As we have seen earlier, LP_{CS} is sound and complete even with respect to single world models (a basic modular model is nothing else than a single world modular model). Unfortunately this does not settle the question of decidability for justification logics at all. The main problem is that a basic evaluation is necessarily not a finite object.

In the following sections we first introduce generated models and then finitary models, which are a subclass of generated models. We show that finitary models satisfy the conditions on the class $\mathbb{C}$ in the previous lemma, which will give us decidability for justification logics. However, we will need certain restrictions on the constant specification in order to obtain a decidable satisfaction relation for finitary models and hence to obtain decidability of LP_{CS}.

4.2 Generated Models

Generated models are models for $\mathsf{LP}_{\mathsf{CS}}$ where the evidence relation is generated by a least fixed point construction. To inductively build-up this least fixed point, we need a monotone operator, which is given as follows.

Definition 4.4 (Evidence closure)**.** Let $\mathcal{B} \subseteq \mathsf{Tm} \times \mathcal{L}_\mathsf{J}$. For a set $X \subseteq \mathsf{Tm} \times \mathcal{L}_\mathsf{J}$ we define $\mathsf{cl}_\mathcal{B}(X)$ by:

1. if $(t, A) \in \mathcal{B}$, then $(t, A) \in \mathsf{cl}_\mathcal{B}(X)$;
2. if $(s, A) \in X$ or $(t, A) \in X$, then $(s + t, A) \in \mathsf{cl}_\mathcal{B}(X)$;
3. if $(s, A) \in X$ and $(t, A \to B) \in X$, then $(t \cdot s, B) \in \mathsf{cl}_\mathcal{B}(X)$;
4. if $(t, A) \in X$, then $(!t, t{:}A) \in \mathsf{cl}_\mathcal{B}(X)$.

Note that $\mathsf{cl}_\mathcal{B}$ is a monotone operator on $\mathsf{Tm} \times \mathcal{L}_\mathsf{J}$, that is

$$X \subseteq Y \quad \text{implies} \quad \mathsf{cl}_\mathcal{B}(X) \subseteq \mathsf{cl}_\mathcal{B}(Y) \tag{4.1}$$

for all $X, Y \subseteq \mathsf{Tm} \times \mathcal{L}_\mathsf{J}$. Hence, as a consequence of the Knaster-Tarski theorem [134], the operator $\mathsf{cl}_\mathcal{B}$ has a least fixed point.

Lemma 4.5 (Least fixed point)**.** *There is a unique $R \subseteq \mathsf{Tm} \times \mathcal{L}_\mathsf{J}$ such that*

1. $\mathsf{cl}_\mathcal{B}(R) = R$,
2. *for any $S \subseteq \mathsf{Tm} \times \mathcal{L}_\mathsf{J}$, if $\mathsf{cl}_\mathcal{B}(S) \subseteq S$, then $R \subseteq S$.*

Proof. Let $C := \{S \subseteq \mathsf{Tm} \times \mathcal{L}_\mathsf{J} \mid \mathsf{cl}_\mathcal{B}(S) \subseteq S\}$. Since $\mathsf{Tm} \times \mathcal{L}_\mathsf{J} \in C$, we know that C is non-empty. Let $R := \bigcap C$. The second claim now holds by definition.

It remains to establish $\mathsf{cl}_\mathcal{B}(R) = R$. Let $S \in C$. Since $R \subseteq S$ and $\mathsf{cl}_\mathcal{B}$ is monotone, we find $\mathsf{cl}_\mathcal{B}(R) \subseteq \mathsf{cl}_\mathcal{B}(S)$. We also have $\mathsf{cl}_\mathcal{B}(S) \subseteq S$, so $\mathsf{cl}_\mathcal{B}(R) \subseteq S$. Since S is an arbitrary element of C and $R = \bigcap C$, this implies $\mathsf{cl}_\mathcal{B}(R) \subseteq R$.

To show $R \subseteq \mathsf{cl}_\mathcal{B}(R)$, we first observe that since $\mathsf{cl}_\mathcal{B}(R) \subseteq R$, we have $\mathsf{cl}_\mathcal{B}(\mathsf{cl}_\mathcal{B}(R)) \subseteq \mathsf{cl}_\mathcal{B}(R)$ by monotonicity. Thus $\mathsf{cl}_\mathcal{B}(R) \in C$, which gives us $R \subseteq \mathsf{cl}_\mathcal{B}(R)$ because $R = \bigcap C$.

Finally, we show the uniqueness of R. Assume that there are two relations $R_1, R_2 \subseteq \mathsf{Tm} \times \mathcal{L}_\mathsf{J}$ both satisfying the two claims of this lemma.

4.2 Generated Models

Then by the first claim for R_1 and the second claim for R_2 we find $R_2 \subseteq R_1$. Similarly we get $R_1 \subseteq R_2$. Hence $R_1 = R_2$ and uniqueness is established. □

Definition 4.6 (Evidence relation). Let $\mathcal{B} \subseteq \mathsf{Tm} \times \mathcal{L}_J$. We define *the minimal evidence relation $\mathcal{E}(\mathcal{B})$ over $\mathcal{B}$* to be the least fixed point of $\mathsf{cl}_\mathcal{B}$.

We have the following property.

Lemma 4.7 (Monotonicity of $\mathcal{E}$). *Let $\mathcal{B}, \mathcal{C} \subseteq \mathsf{Tm} \times \mathcal{L}_J$. We have that*

$$\mathcal{E}(\mathcal{B}) \subseteq \mathcal{E}(\mathcal{B} \cup \mathcal{C}) \ .$$

Proof. By the proof of Lemma 4.5 we know

$$\mathcal{E}(\mathcal{B}) = \bigcap \{X \mid \mathsf{cl}_\mathcal{B}(X) \subseteq X\} \ .$$

Clearly we have $\mathsf{cl}_\mathcal{B}(X) \subseteq \mathsf{cl}_{\mathcal{B} \cup \mathcal{C}}(X)$ for all X. So

$$\mathsf{cl}_{\mathcal{B} \cup \mathcal{C}}(X) \subseteq X \quad \Longrightarrow \quad \mathsf{cl}_\mathcal{B}(X) \subseteq X \ .$$

Hence

$$\mathcal{E}(\mathcal{B}) = \bigcap \{X \mid \mathsf{cl}_\mathcal{B}(X) \subseteq X\} \subseteq \bigcap \{X \mid \mathsf{cl}_{\mathcal{B} \cup \mathcal{C}}(X) \subseteq X\} = \mathcal{E}(\mathcal{B} \cup \mathcal{C})$$

since the first $\bigcap$ is over a larger set. □

Lemma 4.8. *Let $\mathcal{B} \subseteq \mathsf{Tm} \times \mathcal{L}_J$. We have that $\mathcal{E}(\mathcal{E}(\mathcal{B})) = \mathcal{E}(\mathcal{B})$.*

Proof. We have $X \subseteq \mathcal{E}(X)$, so $\supseteq$ holds. By the definition of $\mathsf{cl}_{\mathcal{E}(\mathcal{B})}$ we find $\mathsf{cl}_{\mathcal{E}(\mathcal{B})}(\mathcal{E}(\mathcal{B})) = \mathcal{E}(\mathcal{B})$. So $\mathcal{E}(\mathcal{B})$ is a fixed point of $\mathsf{cl}_{\mathcal{E}(\mathcal{B})}$. Hence, the least fixed point $\mathcal{E}(\mathcal{E}(\mathcal{B}))$ of $\mathsf{cl}_{\mathcal{E}(\mathcal{B})}$ is a subset of $\mathcal{E}(\mathcal{B})$. □

Now we define the class of generated models and the corresponding satisfaction relation.

Definition 4.9 (Generated Model). A *generated model* is a pair

$$\mathcal{M} = (\mathsf{val}, \mathcal{B})$$

where $\mathsf{val} \subseteq \mathsf{Prop}$ and $\mathcal{B} \subseteq \mathsf{Tm} \times \mathcal{L}_J$. For a constant specification CS, the generated model $\mathcal{M}$ is called a *generated CS-model* if $\mathsf{CS} \subseteq \mathcal{B}$.

Definition 4.10. Let $\mathcal{M} = (\mathsf{val}, \mathcal{B})$ be a generated model and D be a formula. We define the relation $\mathcal{M} \Vdash D$ by

1. $\mathcal{M} \not\Vdash \bot$

2. $\mathcal{M} \Vdash P$ iff $P \in \mathsf{val}$

3. $\mathcal{M} \Vdash A \to B$ iff $\mathcal{M} \not\Vdash A$ or $\mathcal{M} \Vdash B$

4. $\mathcal{M} \Vdash t{:}A$ iff $(t, A) \in \mathcal{E}(\mathcal{B})$ and $\mathcal{M} \Vdash A$.

A formula D is *valid with respect to generated* CS*-models* if $\mathcal{M} \Vdash D$ for all generated CS-models $\mathcal{M}$.

Remark 4.11. The above truth definition of $t{:}A$ for generated models is different from that for basic modular models. Basic modular models provide a clear ontological separation of justifications and truth. In generated models, truth and justifications are intertwined in that t is only evidence for A if A is true.

Although ontologically less transparent, generated models provide a very efficient means to establish many important properties of justification logic like decidability.

Theorem 4.12 (Soundness). *For all formulas D,*

$\mathsf{LP}_{\mathsf{CS}} \vdash D$ *implies*   *D is valid with respect to generated* CS*-models.*

Proof. As usual the proof is by induction on the length of the derivation of D. Let $\mathcal{M} = (\mathsf{val}, \mathcal{B})$ be a generated CS-model.

1. D is a propositional tautology. Then D trivially holds in $\mathcal{M}$.

2. D is an instance of **j**, **j+**, or **j4**. In these cases, $\mathcal{M} \Vdash D$ follows immediately from the fact that $\mathcal{E}(\mathcal{B})$ is closed under $\mathsf{cl}_\mathcal{B}$.

3. $D = t{:}A \to A$ is an instance of **jt**. Suppose $\mathcal{M} \Vdash t{:}A$. By Definition 4.10 this implies $\mathcal{M} \Vdash A$ and we are done.

4. D is the conclusion of an instance of (**MP**). It is trivial to see that modus ponens preserves truth in a model.

5. $D = c{:}A$ is the conclusion of an instance of (**AN**$_{\mathsf{CS}}$). Hence

$$(c, A) \in \mathsf{CS} \ .$$

Since $\mathcal{M} = (\mathsf{val}, \mathcal{B})$ is a generated CS-model, we find $(c, A) \in \mathcal{B}$. Since $\mathcal{E}(\mathcal{B})$ is closed under $\mathsf{cl}_\mathcal{B}$, this implies $(c, A) \in \mathcal{E}(\mathcal{B})$. Moreover, since A is an axiom, we have $\mathcal{M} \Vdash A$. Finally we conclude $\mathcal{M} \Vdash c{:}A$.   □

4.2 Generated Models

To establish completeness we again employ a maximal consistent set construction.

Definition 4.13 (Induced model). Let Φ be a maximal $\mathsf{LP}_{\mathsf{CS}}$-consistent set of formulas. The generated model $\mathcal{M}_\Phi = (\mathsf{val}_\Phi, \mathcal{B}_\Phi)$ that is *induced by* Φ is given by

1. $P \in \mathsf{val}_\Phi$ iff $P \in \Phi \cap \mathsf{Prop}$.

2. $(t, A) \in \mathcal{B}_\Phi$ iff $t{:}A \in \Phi$.

$\mathcal{M}_\Phi$ is a generated CS-model. Indeed, suppose $(c, B) \in \mathsf{CS}$. Then we have $\mathsf{LP}_{\mathsf{CS}} \vdash c{:}B$, which by maximal consistency of Φ implies $c{:}B \in \Phi$. We conclude $(c, B) \in \mathcal{B}_\Phi$.

Lemma 4.14. *Let Φ be a maximal $\mathsf{LP}_{\mathsf{CS}}$-consistent set. Then*

$$t{:}A \in \Phi \quad \Longleftrightarrow \quad (t, A) \in \mathcal{E}(\mathcal{B}_\Phi) \ .$$

Proof. We first show

$$\mathsf{cl}_{\mathcal{B}_\Phi}(\mathcal{B}_\Phi) \subseteq \mathcal{B}_\Phi \ . \tag{4.2}$$

We assume $(t, A) \in \mathsf{cl}_{\mathcal{B}_\Phi}(\mathcal{B}_\Phi)$ and distinguish the four cases according to the definition of $\mathsf{cl}_{\mathcal{B}_\Phi}$.

1. $(t, A) \in \mathcal{B}_\Phi$. We are done.

2. $t = t_1 + t_2$ and $(t_1, A) \in \mathcal{B}_\Phi$ or $(t_2, A) \in \mathcal{B}_\Phi$. That means $t_1{:}A \in \Phi$ or $t_2{:}A \in \Phi$. Since Φ is a maximal $\mathsf{LP}_{\mathsf{CS}}$-consistent set, we find that $(t_1 + t_2){:}A \in \Phi$, which yields $(t, A) \in \mathcal{B}_\Phi$.

3. The remaining two cases also follow from the fact that Φ is maximal $\mathsf{LP}_{\mathsf{CS}}$-consistent.

We know by (4.2) that $\mathcal{B}_\Phi$ is a fixed point of $\mathsf{cl}_{\mathcal{B}_\Phi}$. Since $\mathcal{E}(\mathcal{B}_\Phi)$ is the least fixed point of $\mathsf{cl}_{\mathcal{B}_\Phi}$, we find $\mathcal{E}(\mathcal{B}_\Phi) \subseteq \mathcal{B}_\Phi$. Clearly we also have $\mathcal{B}_\Phi \subseteq \mathcal{E}(\mathcal{B}_\Phi)$ and conclude $\mathcal{E}(\mathcal{B}_\Phi) = \mathcal{B}_\Phi$. Now the lemma follows immediately. $\square$

Lemma 4.15 (Truth lemma). *Let Φ be a maximal $\mathsf{LP}_{\mathsf{CS}}$-consistent set of formulas. Then*

$$A \in \Phi \quad \Longleftrightarrow \quad \mathcal{M}_\Phi \Vdash A \ .$$

Proof. By induction on the structure of A.

1. $A \in \mathsf{Prop}$. We have $A \in \Phi$ iff (by definition) $A \in \mathsf{val}_\Phi$ iff (by definition) $\mathcal{M}_\Phi \Vdash A$.

2. $A = \bot$. We have $\bot \notin \Phi$ by maximal consistency of Φ and $\mathcal{M}_\Phi \nVdash \bot$.

3. $A = B \to C$. We have $B \to C \in \Phi$ iff (by maximal consistency of Φ) $B \notin \Phi$ or $C \in \Phi$ iff (by I.H.) $\mathcal{M}_\Phi \nVdash B$ or $\mathcal{M}_\Phi \Vdash C$ iff $\mathcal{M}_\Phi \Vdash B \to C$.

4. $A = t{:}B$. First we note that

$$t{:}B \in \Phi \implies \mathcal{M}_\Phi \Vdash B . \qquad (4.3)$$

Indeed, by **jt** and maximal $\mathsf{LP}_{\mathsf{CS}}$-consistency of Φ we find that $t{:}B \in \Phi$ implies $B \in \Phi$, which implies $\mathcal{M}_\Phi \Vdash B$ by induction hypothesis. So

$$\begin{aligned}
\mathcal{M}_\Phi \Vdash t{:}B &\iff (t,B) \in \mathcal{E}(\mathcal{B}_\Phi) \text{ and } \mathcal{M}_\Phi \Vdash B &\text{by semantics}\\
&\iff t{:}B \in \Phi \text{ and } \mathcal{M}_\Phi \Vdash B &\text{by L. 4.14}\\
&\iff t{:}B \in \Phi &\text{by (4.3)}
\end{aligned}$$

$\square$

As usual, we now obtain completeness of $\mathsf{LP}_{\mathsf{CS}}$ with respect to generated CS-models.

Theorem 4.16 (Completeness). *For all formulas D,*

$$D \text{ is valid with respect to generated } \mathsf{CS}\text{-models} \implies \mathsf{LP}_{\mathsf{CS}} \vdash D .$$

4.3 Finitary models

Let us now introduce a special subclass of generated models, namely the class of *finitary models*, which fulfills the requirements of Lemma 4.3. In this section, we first show that this class is recursively enumerable. Then we establish that the satisfaction relation for finitary models is decidable. In the next section, we prove that $\mathsf{LP}_{\mathsf{CS}}$ is complete with respect to finitary models, which finally gives us decidability for $\mathsf{LP}_{\mathsf{CS}}$.

Definition 4.17 (Finitary model). Let CS be an arbitrary constant specification. Let $\mathcal{C} \subseteq \mathsf{Tm} \times \mathcal{L}_\mathsf{J}$ be finite and set $\mathcal{B} = \mathsf{CS} \cup \mathcal{C}$. Further let val be a finite valuation, that is a finite subset of Prop. Then we call the generated CS-model $\mathcal{M} = (\mathsf{val}, \mathcal{B})$ a *finitary* CS-*model*.

4.3 Finitary models

The finitary CS-model $\mathcal{M} = (\mathsf{val}, \mathcal{B})$ is uniquely determined by the finite set val and the finite set $\mathcal{C}$ with $\mathcal{B} = \mathsf{CS} \cup \mathcal{C}$. Thus we can represent the finitary CS-model $\mathcal{M}$ by the pair $(\mathsf{val}, \mathcal{C})$ and we can recursively enumerate the finitary CS-models in the following sense:

Corollary 4.18. *The class of all pairs* $(\mathsf{val}, \mathcal{C})$ *with* val *being a finite subset of* Prop *and* $\mathcal{C}$ *being a finite subset of* $\mathsf{Tm} \times \mathcal{L}_\mathsf{J}$ *is recursively enumerable.*

In order to show that the satisfaction relation for a finitary model $\mathcal{M} = (\mathsf{val}, \mathcal{B})$ is decidable, we need to decide, in particular, whether $\mathcal{M} \Vdash t{:}F$ holds for a given formula $t{:}F$. That amounts to deciding whether $(t, F) \in \mathcal{E}(\mathcal{B})$ holds.

This, however, cannot be achieved directly. First, note that there are terms that evidence infinitely many formulas. In particular, if CS is schematic and $(c, A) \in \mathsf{CS}$, then c does not only evidence the formula A but also all other (infinitely many) instances of the same axiom scheme.

The problem is to decide whether $(s \cdot t, F) \in \mathcal{E}(\mathcal{B})$. This may hold if there is a formula G with $(s, G \to F) \in \mathcal{E}(\mathcal{B})$ and $(t, G) \in \mathcal{E}(\mathcal{B})$. But we have no bound on the complexity of G here and it does not help that we know the terms s and t because they may evidence infinitely many formulas.

To solve this issue, we need a way to control the set of formulas G such that $(t, G) \in \mathcal{E}(\mathcal{B})$ for a given term t. Therefore, we introduce schematic representations. That means instead of considering infinitely many instances of a schema, we only consider the schema itself so that we can achieve something like

for any given term t, the set $\{F \mid (t, F) \in \mathcal{E}(\mathcal{B})\}$ is finite.

Definition 4.19 (Schematic justification language)**.** We extend the language $\mathcal{L}_\mathsf{J}$ by schematic variables for formulas and schematic variables for terms as follows: Given a term s, for each node ξ in the syntactic tree $\mathfrak{S}$ of s, we add countably many schematic variables Υ_i^ξ for formulas and countably many schematic variables for terms, each indexed by non-negative integers i. We denote the resulting language $\mathcal{L}_\mathsf{J}^s$. Note that $\mathcal{L}_\mathsf{J}$ is a subset of $\mathcal{L}_\mathsf{J}^s$ for any term s. We use $U, V, \ldots$ to denote schemes of formulas, i.e. elements of $\mathcal{L}_\mathsf{J}^s$, as opposed to $F, G, \ldots$ that are reserved for formulas themselves, i.e. elements of $\mathcal{L}_\mathsf{J}$. The letters $r, s, t, \ldots$ may denote both schemes of terms and terms themselves.

For our decidability proof we need substitutions for schematic variables. Hence in this chapter we consider substitutions μ, ν, τ that substitute terms for schematic term variables and formulas for schematic formula variables.

Definition 4.20 (Unifier). Let s be a justification term.

1. Given two $\mathcal{L}_j^s$-schemes U and V, a *unifier* for U and V is a substitution ν such that $U\nu = V\nu$.

2. A unifier μ for U and V is called a *most general unifier* (mgu) if for each unifier ν for U and V, there is a substitution τ such that $\nu = \tau \circ \mu$.

3. An $\mathcal{L}_J$-formula A is called an *instance* of an $\mathcal{L}_j^s$-scheme U, if there is a substitution ν with $A = U\nu$.

A most general unifier need not always exist. However, if it does exist, we will talk of *the* most general unifier. We can do so because if there are several mgus, they are all equivalent up to renaming of schematic variables. For what follows it is important that the existence of a most general unifier is decidable and we can effectively construct one if it exists. We refer to [24] for an introduction to unification.

Remark 4.21. We should explain why we stratify schematic variables over nodes of the syntactic tree of a term. The stratification is necessary for the algorithms we employ for decidability and complexity. The essence of these syntactic devices lies in the difference between subterms and subterm occurrences in a given term. In order to have access to all instances of each scheme we consider, we have to make sure that the schematic representations of different schemes do not interfere with each other's instantiations. If we write $\Upsilon \to (\Upsilon' \to \Upsilon)$ for one axiom scheme and $\Upsilon \to \bot \to \bot \to \Upsilon$ for the other, then we cannot simultaneously instantiate the former to $P \to (Q \to P)$ and the latter to $Q \to \bot \to \bot \to Q$. The superscripts serve as a bookkeeping device for keeping representations of different axiom schemes distinct. This is especially important when we start using unification. But the situation is worse than that. Generally, several schemes may be assigned to one constant. But this constant may have several occurrences in a larger term. The unification computations must mix and match all possible instantiations of these occurrences, which would be impossible if schemes assigned to different occurrences of the same constant have overlaps in

4.3 Finitary models

schematic variables. For instance, in the full LP, all axioms are assigned to all constants. In particular,

$$\mathsf{LP} \vdash (c \cdot c){:}(F \to (G \to F) \to (F \to F)) \tag{4.4}$$

because

$$\mathsf{LP} \vdash c{:}\Big(F \to (G \to F \to F) \to (F \to (G \to F) \to (F \to F))\Big),$$
$$\mathsf{LP} \vdash c{:}(F \to (G \to F \to F))$$

for arbitrary formulas $F, G \in \mathcal{L}_\mathsf{J}$. The proper unification analog of (4.4) should have the form $\Upsilon \to (\Upsilon' \to \Upsilon) \to (\Upsilon \to \Upsilon)$. But consider the overlapping representations of the two axioms:

$$c{:}\Big(\Upsilon_1 \to (\Upsilon_2 \to \Upsilon_3) \to (\Upsilon_1 \to \Upsilon_2 \to (\Upsilon_1 \to \Upsilon_3))\Big),$$
$$c{:}(\Upsilon_2 \to (\Upsilon_1 \to \Upsilon_2))$$

The most general unifier of $\Upsilon_2 \to (\Upsilon_1 \to \Upsilon_2)$ and $\Upsilon_1 \to (\Upsilon_2 \to \Upsilon_3)$ makes $\Upsilon_1 = \Upsilon_2 = \Upsilon_3$, producing the scheme $\Upsilon \to (\Upsilon \to \Upsilon) \to (\Upsilon \to \Upsilon)$ for $c \cdot c$, which does not contain all instances of formulas from (4.4).

Definition 4.22. A constant specification is called *almost schematic* if it is the union of one schematic and one finite constant specification.

Remark 4.23. Let CS be a schematic constant specification and s be a term. It is easy to see that there is a $\mathsf{CS}^s \subseteq \mathsf{Tm} \times \mathcal{L}_\mathsf{J}^s$ such that

1. for each constant c, the set $\{U \mid (c, U) \in \mathsf{CS}^s\}$ is finite;

2. if $(c, A) \in \mathsf{CS}$, then, for each occurrence of c in s, there is a schema U built exclusively from the schematic variables Υ_i^ξ over formulas and schematic variables over terms with the same superscript $\xi \in \mathfrak{S}$ that is the node of the sequent tree of s that contains this occurrence of c, such that $(c, U) \in \mathsf{CS}^s$ and A is an instance of U;

3. if $(c, U) \in \mathsf{CS}^s$, then U is built from schematic variables over formulas and terms with the same superscript ξ corresponding to the same occurrence of c in s and $(c, A) \in \mathsf{CS}$ for all instances A of U.

Let CS be a schematic and CS' be a finite constant specification. Then $\mathsf{CS} \cup \mathsf{CS}'$ is almost schematic and we set $(\mathsf{CS} \cup \mathsf{CS}')^s := \mathsf{CS}^s \cup \mathsf{CS}'$ where CS^s is given as above.

Example 4.24. Let c_1, c_2 be two fixed constants and let P be a fixed atomic proposition. The constant specification

$$\begin{aligned}
\mathsf{CS} = &\{(c_1, F \to (G \to F)) \mid F \text{ and } G \text{ are formulas of } \mathcal{L}_\mathsf{J}\} \cup \\
&\{(c_1, F \to \bot \to \bot \to F) \mid F \text{ is a formula of } \mathcal{L}_\mathsf{J}\} \cup \\
&\{(c_2, P \to (Q_1 \to Q_2) \to (P \to Q_1 \to (P \to Q_2)))\}
\end{aligned}$$

is clearly infinite. Let s be a term such that some node ξ in the syntactic tree $\mathfrak{S}$ of s contains c_1. Then, in the extended language $\mathcal{L}_\mathsf{J}^s$, we can represent it by a finite set

$$\begin{aligned}
\mathsf{CS}^s = &\{(c_1, \Upsilon_1^\xi \to (\Upsilon_2^\xi \to \Upsilon_1^\xi)), (c_1, \Upsilon_3^\xi \to \bot \to \bot \to \Upsilon_3^\xi), \\
&(c_2, P \to (Q_1 \to Q_2) \to (P \to Q_1 \to (P \to Q_2)))\} \ .
\end{aligned}$$

We will establish decidability of LP_CS only for constant specifications that are both decidable and almost schematic. Later, in Theorem 4.37 we show that for arbitrary constant specifications decidability can fail.

Definition 4.25. Let s be a justification term. Let r_ξ denote the term in a node ξ of the syntactic tree $\mathfrak{S}$ of s. We assume that for each such node $\xi \in \mathfrak{S}$ we are given a finite set $\|\xi\|^0 \subseteq \mathcal{L}_\mathsf{J}^s$ of schemes with all variables over formulas and over terms occurring in it having the same superscript ξ. For each positive natural number i and each node $\xi \in \mathfrak{S}$ we inductively define a set $\|\xi\|^i \subseteq \mathcal{L}_\mathsf{J}^s$ as follows:

1. If r_ξ is atomic, then $\|\xi\|^{i+1} := \|\xi\|^i$.

2. Let $r_\xi = r_\kappa + r_\chi$ for the children κ and χ of ξ. Then
$$\|\xi\|^{i+1} := \|\xi\|^i \cup \|\kappa\|^i \cup \|\chi\|^i \ .$$

3. Let $r_\xi = r_\kappa \cdot r_\chi$ for the children κ and χ of ξ. Then $\|\xi\|^{i+1}$ is the smallest superset of $\|\xi\|^i$ such that

    (a) if $\Upsilon_j^{\kappa'} \in \|\kappa\|^i$ for some descendant κ' of κ in $\mathfrak{S}$ and for some schematic variable $\Upsilon_j^{\kappa'}$ over formulas and if $\|\chi\|^i \neq \varnothing$, then $\Upsilon_0^\xi \in \|\xi\|^{i+1}$;[2]

---

[2]This case may seem unnecessary and overly complicated at the moment but we will use it in the proof of Lemma 4.30, as well as in the next chapter. Also note the special role of subscript 0, which is reserved to handle this case. In particular, in the proof of Lemma 4.30, we will require that schematic variables with subscript 0 may not occur in CS^s.

4.3 Finitary models

(b) if there exist $U_1, U_2, V_1 \in \mathcal{L}_j^s$, and ν with
   i. $U_1 \to V_1 \in \|\kappa\|^i$,
   ii. $U_2 \in \|\chi\|^i$,
   iii. ν is the most general unifier of U_1 and U_2, computed using a fixed deterministic algorithm,[3]

   then $V_1\nu \in \|\xi\|^{i+1}$.

4. Let $r_\xi = !r_\chi$ for the child χ of ξ. Then
$$\|\xi\|^{i+1} := \|\xi\|^i \cup \{r_\chi{:}U \mid U \in \|\chi\|^i\} \ .$$

Example 4.26. Let $s = !((x \cdot y) \cdot z)$. The syntactic tree $\mathfrak{S}$ of s consists of six nodes, $\rho, \xi, \chi, \chi_1, \chi_2,$ and χ_3 where ρ is the root, ξ is its only child, χ and χ_3 are the two children of ξ, and χ_1 and χ_2 are the two children of χ. In particular,
$$r_\rho = s = !((x \cdot y) \cdot z), \quad r_\xi = (x \cdot y) \cdot z, \quad r_\chi = x \cdot y,$$
$$r_{\chi_1} = x, \quad r_{\chi_2} = y, \quad r_{\chi_3} = z \ ,$$
see Figure 4.1.

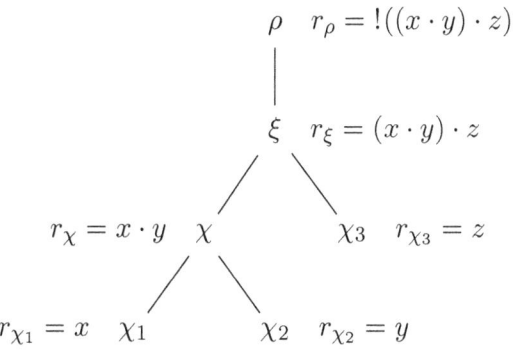

Figure 4.1: Term structure

Let
$$\|\rho\|^0 = \varnothing, \quad \|\xi\|^0 = \{P \wedge Q\}, \quad \|\chi\|^0 = \{\Upsilon_0^\chi \to (Q \to \Upsilon_0^\chi)\},$$
$$\|\chi_1\|^0 = \{(\Upsilon_1^{\chi_1} \to \Upsilon_1^{\chi_1}) \to \Upsilon_3^{\chi_1}\}, \quad \|\chi_2\|^0 = \{Q \to Q\}, \quad \|\chi_3\|^0 = \{P\} \ .$$
Applying Definition 4.25, we compute

---
[3] Note that ν does not introduce any new variables over formulas or terms.

1. $\|\chi_1\|^i = \|\chi_1\|^0 = \{(\Upsilon_1^{\chi_1} \to \Upsilon_1^{\chi_1}) \to \Upsilon_3^{\chi_1}\}$ for all $i \geq 0$;
2. $\|\chi_2\|^i = \|\chi_2\|^0 = \{Q \to Q\}$ for all $i \geq 0$;
3. $\|\chi_3\|^i = \|\chi_3\|^0 = \{P\}$ for all $i \geq 0$;
4. $\|\chi\|^1 = \{\Upsilon_3^{\chi_1}, \quad \Upsilon_0^{\chi} \to (Q \to \Upsilon_0^{\chi})\}$ because the mgu of

$$\Upsilon_1^{\chi_1} \to \Upsilon_1^{\chi_1} \quad \text{and} \quad Q \to Q$$

is $\nu_1 := [\Upsilon_1^{\chi_1} \mapsto Q]$ and $\Upsilon_3^{\chi_1}\nu_1 = \Upsilon_3^{\chi_1}$;

5. $\|\xi\|^1 = \{Q \to P, \quad P \wedge Q\}$ because the mgu of Υ_0^{χ} and P is

$$\nu_2 := [\Upsilon_0^{\chi} \mapsto P]$$

and $(Q \to \Upsilon_0^{\chi})\nu_2 = Q \to P$;

6. $\|\rho\|^1 = \{((x \cdot y) \cdot z){:}(P \wedge Q)\}$;
7. $\|\chi\|^i = \|\chi\|^1 = \{\Upsilon_3^{\chi_1}, \quad \Upsilon_0^{\chi} \to (Q \to \Upsilon_0^{\chi})\}$ for all $i \geq 1$;
8. $\|\xi\|^2 = \{\Upsilon_0^{\xi}, \quad Q \to P, \quad P \wedge Q\}$;
9. $\|\rho\|^2 = \{((x \cdot y) \cdot z){:}(Q \to P), \quad ((x \cdot y) \cdot z){:}(P \wedge Q)\}$;
10. $\|\xi\|^i = \|\xi\|^2 = \{\Upsilon_0^{\xi}, \quad Q \to P, \quad P \wedge Q\}$ for all $i \geq 2$;
11. $\|\rho\|^3 = \{((x\cdot y)\cdot z){:}\Upsilon_0^{\xi}, \quad ((x\cdot y)\cdot z){:}(Q \to P), \quad ((x\cdot y)\cdot z){:}(P \wedge Q)\}$;
12. $\|\rho\|^i = \|\rho\|^3$ for all $i \geq 3$.

Lemma 4.27. *Let s be a term and $\mathfrak{S}$ be the syntactic tree of s and let $\|\xi\|^0 \subseteq \mathcal{L}_j^s$ be finite for each $\xi \in \mathfrak{S}$. Then, for each natural number i and each node $\xi \in \mathfrak{S}$, the set $\|\xi\|^i$ is finite and can be effectively constructed. Each variable over formulas or terms occurring in $\|\xi\|^i$ has superscript κ for some descendant κ of ξ in $\mathfrak{S}$.*

Proof. By induction on i. For $i = 0$ the claim holds by assumption. To show that $\|\xi\|^{i+1}$ is finite and only contains schematic variables from ξ's descendants, we first observe that by I.H. the claim holds for $\|\xi\|^i$. Thus it is enough to show that only finitely many formulas are added, they can be effectively constructed, and the only new variable over formulas that can be introduced corresponds to ξ itself. We distinguish the cases for r_ξ.

4.3 Finitary models

1. $r_\xi = r_\kappa + r_\chi$. The claim follows by I.H. for $\|\kappa\|^i$ and $\|\chi\|^i$.

2. $r_\xi = r_\kappa \cdot r_\chi$. We apply I.H. for $\|\kappa\|^i$ and $\|\chi\|^i$. There are two cases:

    (a) It is possible to effectively verify whether $\|\kappa\|^i$ contains a variable over formulas among its finitely many members. If so it is also possible to check whether $\|\chi\|^i$ is not empty. If both conditions are satisfied, Υ_0^ξ is added to $\|\xi\|^{i+1}$. It remains to note that this variable over formulas corresponds to ξ, which is a (trivial) descendant of ξ in $\mathfrak{S}$.

    (b) For each implication $U_1 \to V_1 \in \|\kappa\|^i$, of which there are finitely many, and for each of the finitely many $U_2 \in \|\chi\|^i$, it is possible to effectively compute the mgu ν of U_1 and U_2, using our fixed deterministic unification algorithm. Further, it is possible to effectively construct $V = V_1\nu$. Moreover these V can also be effectively constructed. Since the unification process does not introduce new variables over formulas or terms, all such variables in V must stem from U_1, U_2, or V_1, i.e., they must correspond to descendants of κ and χ, the children of ξ in $\mathfrak{S}$.

3. $r_\xi = !r_\chi$. By I.H. the set $\|\chi\|^i$ is finite and it can be effectively constructed. It is sufficient to add all its elements prefixed with "r_χ:" to $\|\xi\|^{i+1}$. It is obvious that no new variables over formulas or terms is introduced in the process. □

As we saw in the example, the inductive construction of $\|\xi\|^i$ reaches a fixed point for some finite i, that is there is an i such that $\|\xi\|^j = \|\xi\|^i$ for all $j \geq i$. We show that this natural number i is bounded by the complexity of the term r_ξ, where the complexity of a term is measured by a rank function.

Definition 4.28. The *rank* $\mathsf{rk}(t)$ of a term t is inductively defined by:

1. $\mathsf{rk}(c) := \mathsf{rk}(x) := 0$

2. $\mathsf{rk}(s+t) := \max(\mathsf{rk}(s), \mathsf{rk}(t)) + 1$

3. $\mathsf{rk}(s \cdot t) := \max(\mathsf{rk}(s), \mathsf{rk}(t)) + 1$

4. $\mathsf{rk}(!s) := \mathsf{rk}(s) + 1$

Lemma 4.29. *Let s be a justification term and $\mathfrak{S}$ be the syntactic tree of s. Let $\|\xi\|^0 \subseteq \mathcal{L}_j^s$ be finite for each $\xi \in \mathfrak{S}$. For each natural number i and each node $\xi \in \mathfrak{S}$ we have that for all $U \in \mathcal{L}_j^s$*

$$U \in \|\xi\|^i \quad \text{implies} \quad U \in \|\xi\|^{\mathsf{rk}(r_\xi)} \ .$$

Proof. First remember that we have

$$k \leq l \quad \text{implies} \quad \|\xi\|^k \subseteq \|\xi\|^l \ . \tag{4.5}$$

We show the lemma by induction on i. For $i = 0$ it is an immediate consequence of (4.5). For the induction step suppose $U \in \|t\|^{i+1}$. If $U \in \|t\|^i$, then $U \in \|t\|^{\mathsf{rk}(t)}$ follows immediately by I.H. Otherwise we have one of the following cases.

1. $r_\xi = r_\kappa + r_\chi$ and $U \in \|\kappa\|^i \cup \|\chi\|^i$. The claim follows from the I.H. and $\mathsf{rk}(r_\kappa), \mathsf{rk}(r_\chi) < \mathsf{rk}(r_\xi)$.

2. $r_\xi = r_\kappa \cdot r_\chi$. Let $j := \mathsf{rk}(r_\xi) - 1 \geq \mathsf{rk}(r_\kappa), \mathsf{rk}(r_\chi)$. We distinguish two cases.

   (a) There exists $\Upsilon_l^{\kappa'} \in \|\kappa\|^i$ for some descendant κ' of κ in $\mathfrak{S}$ and $\|\chi\|^i \neq \varnothing$. By I.H., $\Upsilon_l^{\kappa'} \in \|\kappa\|^{\mathsf{rk}(r_\kappa)}$ and $\|\chi\|^{\mathsf{rk}(r_\chi)} \neq \varnothing$. Thus, by (4.5), $\Upsilon_l^{\kappa'} \in \|\kappa\|^j$ and $\|\chi\|^j \neq \varnothing$. Thus,

   $$U = \Upsilon_0^\xi \in \|\xi\|^{\mathsf{rk}(r_\xi)} \ ,$$

   as required.

   (b) There are V_1, V_2, U_1 with $V_1 \to U_1 \in \|\kappa\|^i$ and $V_2 \in \|\chi\|^i$ and $U = U_1\nu$ where ν is the most general unifier of V_1 and V_2 computed by our fixed deterministic unification algorithm. As above, we find $V_1 \to U_1 \in \|\kappa\|^j$ and $V_2 \in \|\chi\|^j$. Hence we obtain $U \in \|\xi\|^{\mathsf{rk}(r_\xi)}$.

3. $r_\xi = !r_\kappa$. We have $U = r_\kappa{:}V$ and $V \in \|\kappa\|^i$. By I.H. we get

$$V \in \|\kappa\|^{\mathsf{rk}(r_\kappa)}$$

and thus $r_\kappa{:}V \in \|\xi\|^{\mathsf{rk}(r_\kappa)+1}$, which is $U \in \|\xi\|^{\mathsf{rk}(r_\xi)}$. $\square$

Now we can show that the evidence relation $\mathcal{E}(\mathcal{B})$ for finitary models is decidable.

4.3 Finitary models

Lemma 4.30. *Let* CS *be a decidable and almost schematic constant specification. Let* $\mathcal{X} \subseteq \mathsf{Tm} \times \mathcal{L}_\mathsf{J}$ *be finite. Set* $\mathcal{B} = \mathsf{CS} \cup \mathcal{X}$.
For each term s and each formula G, it is decidable whether

$$(s, G) \in \mathcal{E}(\mathcal{B}) \ .$$

Proof. Let CS^s constructed using schemes from $\mathcal{L}_\mathsf{J}^s$ be the finite representation of CS constructed as in Remark 4.23 in such a way that no two distinct elements of CS^s use the same variable over formulas or the same variable over terms[4] and that no element of CS^s uses a variable over formulas with subscript 0. Let $\mathfrak{S}$ be the syntactic tree of the term s. First we construct $\|\xi\|^0$ for each node ξ of $\mathfrak{S}$ as follows:

$$\|\xi\|^0 := \{U \mid (r_\xi, U) \in \mathsf{CS}^s\} \cup \{F \mid (r_\xi, F) \in \mathcal{X}\} \ .$$

It is clear that $\|\xi\|^0$ is finite for each $\xi \in \mathfrak{S}$. Now we show for each node $\xi \in \mathfrak{S}$ that

$(r_\xi, E) \in \mathcal{E}(\mathcal{B})$ if and only if

there exists i such that E is an instance of some $U \in \|\xi\|^i$.   (4.6)

We show the direction from left to right by induction on the depth of ξ in $\mathfrak{S}$.

1. ξ is a leaf of $\mathfrak{S}$, i.e., r_ξ is atomic. $(r_\xi, E) \in \mathcal{E}(\mathcal{B})$ implies $(r_\xi, E) \in \mathcal{B}$. If $(r_\xi, E) \in \mathcal{X}$, then $E \in \|\xi\|^0$ and we are done. Otherwise, $(r_\xi, E) \in \mathsf{CS}$, which implies that E is an instance of some U with $U \in \|\xi\|^0$.

2. $r_\xi = r_\kappa + r_\chi$ for the children κ and χ of ξ in $\mathfrak{S}$. If $(r_\xi, E) \in \mathcal{B}$, we proceed as in the first case. Otherwise, suppose $(r_\kappa, E) \in \mathcal{E}(\mathcal{B})$. By I.H. we find that E is an instance of some $U \in \|\kappa\|^i$ for some i. The claim follows by $\|\kappa\|^i \subseteq \|\xi\|^{i+1}$. The case for $(r_\chi, E) \in \mathcal{E}(\mathcal{B})$ is similar.

3. $r_\xi = r_\kappa \cdot r_\chi$ for the children κ and χ of ξ in $\mathfrak{S}$. If $(r_\xi, E) \in \mathcal{B}$, we proceed as in the first case. Otherwise, there is a formula $H \in \mathcal{L}_\mathsf{J}$ such that

$$(r_\kappa, H \to E) \in \mathcal{E}(\mathcal{B}) \quad \text{and} \quad (r_\chi, H) \in \mathcal{E}(\mathcal{B}) \ .$$

---
[4]This property is automatically guaranteed for elements corresponding to distinct constant occurrences in s. However, CS may assign several axiom schemes to the same constant, i.e., to all occurrences of this constant in s. It is necessary to make sure that these axiom schemes do not interfere with each other's unification.

By I.H. we find that there are $V_2 \in \mathcal{L}_j^s$ and i_v such that

$$V_2 \in \|\chi\|^{i_v} \quad \text{and} \quad H = V_2 \tau_\chi \text{ for some substitution } \tau_\chi \ .$$

Also by I.H. there are $W \in \mathcal{L}_j^s$ and i_u such that

$$W \in \|\kappa\|^{i_u} \quad \text{and} \quad H \to E = W \tau_\kappa \text{ for some substitution } \tau_\kappa \ .$$

Now we distinguish two cases.

(a) $W = \Upsilon_l^{\kappa'}$ for some descendant κ' of κ and some natural number l. Then, by Definition 4.25, since $\|\chi\|^{\max(i_u, i_v)} \neq \emptyset$ because it contains V_2,

$$\Upsilon_0^\xi \in \|\xi\|^{\max(i_u, i_v)+1}$$

and, clearly, E is an instance of this Υ_0^ξ.

(b) W has the form $V_1 \to U_1$. Then

$$H = V_1 \tau_\kappa \quad \text{and} \quad E = U_1 \tau_\kappa \ .$$

All variables over formulas and over terms in $V_1 \to U_1$ correspond to nodes in the subtree rooted at κ, whereas all those in V_2 correspond to nodes in a disjoint subtree rooted at χ. Thus, there are no common variables over formulas or over terms between $V_1 \to U_1$ and V_2 and, w.l.o.g., we may assume that τ_κ and τ_χ are each defined only on the variables occurring in $V_1 \to U_1$ and in V_2 respectively. Therefore, the joint substitution $\tau_\kappa \sqcup \tau_\chi$ is defined and

$$(V_1 \to U_1)(\tau_\kappa \sqcup \tau_\chi) = (V_1 \to U_1)\tau_\kappa = H \to E \quad \text{and}$$
$$V_2(\tau_\kappa \sqcup \tau_\chi) = V_2 \tau_\chi = H \ .$$

In other words, $\tau_\kappa \sqcup \tau_\chi$ is a unifier of V_1 and V_2. Let ν be the most general unifier of V_1 and V_2. We find that there is a substitution μ such that $\tau_\kappa \sqcup \tau_\chi = \mu \circ \nu$, Hence

$$E = U_1(\mu \circ \nu)$$

and thus E is an instance of $U := U_1 \nu$. Moreover

$$U \in \|\xi\|^{\max(i_u, i_v)+1} \ .$$

4.3 Finitary models

4. $r_\xi = !r_\chi$ for the child χ of ξ in $\mathfrak{G}$. If $(r_\xi, E) \in \mathcal{B}$, we proceed as in the first case. Otherwise there is a formula H such that $E = r_\chi{:}H$ and $(r_\chi, H) \in \mathcal{E}(\mathcal{B})$. By I.H. there are $V \in \mathcal{L}_j^s$ and i such that

$$V \in \|\chi\|^i \quad \text{and} \quad H \text{ is an instance of } V \ .$$

Thus we get $r_\chi{:}V \in \|\xi\|^{i+1}$ and $r_\chi{:}H$ is an instance of $r_\chi{:}V$.

We show the direction from right to left of (4.6) by induction on i.

Base case $i = 0$. There are two possible subcases: either E is a trivial instance of itself and $(r_\xi, E) \in \mathcal{X}$ or E is an instance of an axiom scheme U such that $(r_\xi, U) \in \mathsf{CS}^s$ and $(r_\xi, E) \in \mathsf{CS}$. In both cases we find $(r_\xi, E) \in \mathcal{B}$ and thus $(r_\xi, E) \in \mathcal{E}(\mathcal{B})$.

Induction step $i = j + 1$. If E is an instance of some $U \in \|\xi\|^j$, the claim follows by I.H. Otherwise we distinguish the following cases.

1. $r_\xi = r_\kappa + r_\chi$ for the children κ and χ of ξ in $\mathfrak{G}$ and E is an instance of some $U \in \|\kappa\|^j$. By I.H. we get $(r_\kappa, E) \in \mathcal{E}(\mathcal{B})$. By the closure conditions on $\mathcal{E}(\mathcal{B})$, we find $(r_\kappa + r_\chi, E) \in \mathcal{E}(\mathcal{B})$.

2. The case for $r_\xi = r_\kappa + r_\chi$ and E being an instance of some $U \in \|\chi\|^j$ is analogous.

3. $r_\xi = r_\kappa \cdot r_\chi$ for the children κ and χ of ξ in $\mathfrak{G}$. There are two cases.

   (a) There exist a variable over formulas $\Upsilon_l^{\kappa'} \in \|\kappa\|^j$ and

   $$U_2 \in \|\chi\|^j \ .$$

   Let $H \in \mathcal{L}_j$ be an instance of U_2. Clearly, $H \to E$ is an instance of $\Upsilon_l^{\kappa'}$. By I.H. we find

   $$(r_\chi, H) \in \mathcal{E}(\mathcal{B}) \quad \text{and} \quad (r_\kappa, H \to E) \in \mathcal{E}(\mathcal{B}) \ .$$

   We conclude $(r_\kappa \cdot r_\chi, E) \in \mathcal{E}(\mathcal{B})$.

   (b) There are V_1, V_2, U and ν such that

   i. $V_1 \to U \in \|\kappa\|^j$,

   ii. $V_2 \in \|\chi\|^j$,

   iii. ν is the mgu of V_1 and V_2, and

   iv. $E = (U\nu)\mu$ for some substitution μ.

Since ν is a unifier for V_1 and V_2, $(V_1\nu)\mu = (V_2\nu)\mu$. If there are still schematic variables left in $(V_1\nu)\mu = (V_2\nu)\mu$, a further substitution μ' can yield $H = ((V_1\nu)\mu)\mu' = ((V_2\nu)\mu)\mu' \in \mathcal{L}_J$. Since $E = (U\nu)\mu$ is already an $\mathcal{L}_J$-formula, it follows that

$$E = (U\nu)\mu = ((U\nu)\mu)\mu' \ .$$

Thus, $H \to E = (V_1 \to U)(\mu' \circ \mu \circ \nu)$ is an instance of $V_1 \to U$. By I.H. we find $(r_\kappa, H \to E) \in \mathcal{E}(\mathcal{B})$. Moreover, since

$$H = V_2(\mu' \circ \mu \circ \nu)$$

is an instance of V_2, we also find by I.H. that $(r_\chi, H) \in \mathcal{E}(\mathcal{B})$. Hence by the closure conditions on $\mathcal{E}(\mathcal{B})$, we conclude that

$$(r_\kappa \cdot r_\chi, E) \in \mathcal{E}(\mathcal{B}) \ .$$

4. $r_\xi = !r_\chi$ for the child χ of ξ in $\mathfrak{S}$ and E is an instance of $U = r_\chi{:}V$ for some $V \in \|\chi\|^j$. Then E has the form $r_\chi{:}H$ for some instance H of V. By I.H. we find $(r_\chi, H) \in \mathcal{E}(\mathcal{B})$ and thus $(!r_\chi, r_\chi{:}H) \in \mathcal{E}(\mathcal{B})$, which is $(r_\xi, E) \in \mathcal{E}(\mathcal{B})$.

Hence (4.6) is established. By Lemma 4.29 we obtain

$(r_\xi, E) \in \mathcal{E}(\mathcal{B})$ if and only if

E is an instance of some $U \in \|\xi\|^{\mathsf{rk}(r_\xi)}$. (4.7)

Now we can easily decide $(s, G) \in \mathcal{E}(\mathcal{B})$ as follows. Construct $\|\rho\|^{\mathsf{rk}(r_\rho)}$ for the root ρ of $\mathfrak{S}$, where $r_\rho = s$. We have shown that this is a finite set. For each $U \in \mathcal{L}_J^s$ in this set, decide whether G is an instance of U. By (4.7) we have $(s, G) \in \mathcal{E}(\mathcal{B})$ if and only if such a U exists. □

Theorem 4.31. *Let* CS *be a decidable and almost schematic constant specification. The satisfaction relation for finitary* CS*-models is decidable.*

Proof. Let $\mathcal{X} \subseteq \mathsf{Tm} \times \mathcal{L}_J$ be finite and set $\mathcal{B} = \mathsf{CS} \cup \mathcal{X}$. Further let val be a finite valuation. We consider the finitary model $\mathcal{M} = (\mathsf{val}, \mathcal{B})$ and show that for any formula F, it is decidable whether $\mathcal{M} \Vdash F$.

The proof is by induction on F. We distinguish the following cases.

1. F is an atomic proposition P. Since val is finite, it is decidable whether $P \in \mathsf{val}$, that is whether $\mathcal{M} \Vdash F$.

2. $F = \bot$. We have $\mathcal{M} \not\Vdash \bot$. Thus $\mathcal{M} \Vdash F$ trivially is decidable.

3. $F = G \to H$. By I.H. $\mathcal{M} \Vdash G$ and $\mathcal{M} \Vdash H$ are decidable and thus $\mathcal{M} \Vdash F$ is decidable, too.

4. $F = s{:}G$. By the previous lemma $(s, G) \in \mathcal{E}(\mathcal{B})$ is decidable and by I.H. $\mathcal{M} \Vdash G$ also is decidable. Thus $\mathcal{M} \Vdash F$ is decidable, too. □

4.4 Establishing decidability

To establish decidability of $\mathsf{LP}_{\mathsf{CS}}$ for decidable and almost schematic constant specifications CS, it remains to show that $\mathsf{LP}_{\mathsf{CS}}$ is complete with respect to finitary CS-models. We know that $\mathsf{LP}_{\mathsf{CS}}$ is complete for generated CS-models. Hence if F is not derivable in $\mathsf{LP}_{\mathsf{CS}}$, there exists a generated CS-model $\mathcal{M}$ that falsifies F. In this section, we show that we can restrict $\mathcal{M}$ to a finitary CS-model $\mathcal{M} \restriction F$ such that $\mathcal{M} \restriction F \not\Vdash F$.

Definition 4.32 (Restricted model)**.** Let CS be a constant specification and $\mathcal{M} = (\mathsf{val}, \mathcal{B})$ be a generated CS-model. Further let F be a formula. We define the restricted generated CS-model $\mathcal{M} \restriction F = (\mathsf{val} \restriction F, \mathcal{B} \restriction F)$ by

1. $\mathsf{val} \restriction F := \{P \mid P \in \mathsf{val} \text{ and } P \in \mathsf{sub}(F)\}$

2. $\mathcal{B} \restriction F := \{(t, A) \mid (t, A) \in \mathcal{E}(\mathcal{B}) \text{ and } t{:}A \in \mathsf{sub}(F)\} \cup \mathsf{CS}$.

Remark 4.33. Let CS be an almost schematic constant specification and $\mathcal{M} = (\mathsf{val}, \mathcal{B})$ be a generated CS-model. Since there are only finitely many subformulas of a given formula F, the set $\mathsf{sub}(F)$ is finite. Therefore, the restricted generated CS-model $\mathcal{M} \restriction F$ is finitary.

Lemma 4.34. *Let* $\mathcal{M} = (\mathsf{val}, \mathcal{B})$ *be a generated* CS*-model and* F *a formula. For all* $A \in \mathsf{sub}(F)$ *we have*

$$\mathcal{M} \Vdash A \quad \textit{if and only if} \quad \mathcal{M} \restriction F \Vdash A \ .$$

Proof. First we show that for all $t{:}B \in \mathsf{sub}(F)$ we have

$$(t, B) \in \mathcal{E}(\mathcal{B}) \quad \textit{if and only if} \quad (t, B) \in \mathcal{E}(\mathcal{B} \restriction F) \ . \qquad (4.8)$$

From left to right: $(t, B) \in \mathcal{E}(\mathcal{B})$ and $t{:}B \in \mathsf{sub}(F)$ imply $(t, B) \in \mathcal{B} \restriction F$ and hence $(t, B) \in \mathcal{E}(\mathcal{B} \restriction F)$. From right to left: Observe $\mathsf{CS} \subseteq \mathcal{B} \subseteq \mathcal{E}(\mathcal{B})$ since $\mathcal{M}$ is a CS-model. So, $\mathsf{CS} \subseteq \mathcal{E}(\mathcal{B})$. Therefore, $\mathcal{B} \restriction F \subseteq \mathcal{E}(\mathcal{B})$ by definition of $\mathcal{B} \restriction F$. By Lemma 4.7 this implies $\mathcal{E}(\mathcal{B} \restriction F) \subseteq \mathcal{E}(\mathcal{E}(\mathcal{B}))$. By

Lemma 4.8 we know $\mathcal{E}(\mathcal{E}(\mathcal{B})) = \mathcal{E}(\mathcal{B})$, which yields $\mathcal{E}(\mathcal{B} \restriction F) \subseteq \mathcal{E}(\mathcal{B})$. Hence (4.8) is established for all $t{:}B \in \mathsf{sub}(F)$.

Now the claim of our lemme, i.e.,

$$\mathcal{M} \Vdash A \quad \text{if and only if} \quad \mathcal{M} \restriction F \Vdash A \quad \text{for all } A \in \mathsf{sub}(F),$$

easily follows by induction on A. We distinguish the following cases.

1. $A = P$ for $P \in \mathsf{Prop}$. We have $\mathcal{M} \Vdash P$ iff $P \in \mathsf{val}$ iff (since $P \in \mathsf{sub}(F)$) $P \in \mathsf{val} \restriction F$ iff $\mathcal{M} \restriction F \Vdash P$.

2. $A = \bot$. We have $\mathcal{M} \not\Vdash A$ and $\mathcal{M} \restriction F \not\Vdash A$.

3. $A = B \to C$. This case follows immediately by I.H.

4. $A = t{:}B$. We have $\mathcal{M} \Vdash t{:}B$ if and only if

$$(t, B) \in \mathcal{E}(\mathcal{B}) \quad \text{and} \quad \mathcal{M} \Vdash B \ .$$

Because of $t{:}B \in \mathsf{sub}(F)$, (4.8), $B \in \mathsf{sub}(F)$, and I.H., this is equivalent to

$$(t, B) \in \mathcal{E}(\mathcal{B} \restriction F) \quad \text{and} \quad \mathcal{M} \restriction F \Vdash B \ ,$$

which is equivalent to $\mathcal{M} \restriction F \Vdash t{:}B$. □

Now we can show completeness with respect to finitary models.

Theorem 4.35. *Let* CS *be an almost schematic constant specification. Let F be a formula that is not derivable in* $\mathsf{LP}_{\mathsf{CS}}$. *Then there exists a finitary* CS*-model* $\mathcal{M}_{\mathit{fin}}$ *with* $\mathcal{M}_{\mathit{fin}} \not\Vdash F$.

Proof. Let F be such that $\mathsf{LP}_{\mathsf{CS}} \not\vdash F$. By Theorem 4.16 there is a generated model $\mathcal{M}$ such that $\mathcal{M} \not\Vdash F$. Consider the restricted model $\mathcal{M} \restriction F$. By Remark 4.33 we know that $\mathcal{M} \restriction F$ is a finitary CS-model. Moreover, by Lemma 4.34 we conclude $\mathcal{M} \restriction F \not\Vdash F$. □

Corollary 4.36. *Let* CS *be a decidable and almost schematic constant specification. Then* $\mathsf{LP}_{\mathsf{CS}}$ *is decidable.*

Proof. Let $\mathbb{C}$ be the class of finitary CS-models for a decidable and almost schematic constant specification CS. By Theorem 4.12 we know that $\mathsf{LP}_{\mathsf{CS}}$ is sound with respect to $\mathbb{C}$ and Theorem 4.35 gives us completeness of $\mathsf{LP}_{\mathsf{CS}}$ with respect to $\mathbb{C}$. The class $\mathbb{C}$ is recursively enumerable by Corollary 4.18. Finally, by Theorem 4.31, the binary relation $\mathcal{M} \Vdash F$ between formulas and models from $\mathbb{C}$ is decidable. Thus we have established the assumptions of Lemma 4.3 and conclude that $\mathsf{LP}_{\mathsf{CS}}$ is decidable. □

4.5 A decidable constant specification is not enough

The requirement for the constant specification to be almost schematic cannot be dropped completely from Corollary 4.36.

Theorem 4.37. *There exists a decidable constant specification* CS *such that* $\mathsf{LP}_{\mathsf{CS}}$ *is undecidable.*

Proof. The proof is by reducing the Halting Problem to provability in $\mathsf{LP}_{\mathsf{CS}}$ for a particular CS. Let T_i stand for the ith Turing machine with one input. Let $A_0, A_1, \ldots$ be an effective enumeration of all axioms of LP. Consider the following constant specification

$$\mathsf{CS} := \big\{ \big(c, (A_i \to (A_j \to A_i))\big) \mid T_i(i) \text{ halts after at most } j \text{ steps} \big\}$$
$$\cup \, \{(d, A_i) \mid i = 0, 1, \ldots\} \ .$$

Clearly this constant specification is decidable. At the same time, it can easily be shown that

$$\mathsf{LP}_{\mathsf{CS}} \vdash ((c \cdot d) \cdot d){:}A_i \quad \text{if and only if} \quad T_i(i) \text{ halts} \ . \tag{4.9}$$

The right side of this equivalence is the Halting Problem, which is known to be undecidable. □

Whether this requirement of almost schematicness can be significantly weakened is not known.

4.6 Notes

Artemov established decidability for $\mathsf{LP}_{\mathsf{CS}}$ with a finite constant specification CS already in [4]. Later, Mkrtychev [114] showed that $\mathsf{LP}_{\mathsf{CS}}$ is decidable when CS is a schematic constant specification. Kuznets [90] provided decidability results for justification logics with schematic constant specifications that correspond to the modal logics K, KT, and K4. Studer [132] established decidability for a justification logic that corresponds to the modal logic S5 where the constant specification is again required to be finite.

Bucheli, Kuznets, and Studer [37] adapt model-theoretic tools, in particular filtrations and a certain form of generated submodel, to the context of justification logic. This allows them to give another proof of the decidability of $\mathsf{LP_{CS}}$.

Theorem 4.37 about the undecidability of $\mathsf{LP_{CS}}$ for arbitrary constant specifications goes back to Kuznets [91].

The notion of an *almost schematic* CS is also due to Kuznets [92].

Chapter 5

Complexity

In the preceding chapter, we have shown that the Logic of Proofs $\mathsf{LP}_{\mathsf{CS}}$ is decidable for all almost schematic constant specifications CS. However, the complexity of the decision procedure differ depending on CS. This chapter is devoted to the results about this complexity and a comparison with the complexity of such algorithms for S4 and the intuitionistic logic.

5.1 Short Introduction to Complexity Classes

The complexity of a logic L usually means the complexity of the *validity problem* for L, i.e., of the problem, given a formula A, to determine whether $\mathsf{L} \vdash A$. If the logic L is supplied with a semantics, another problem is often considered: the *satisfiability problem* for L, i.e., given a formula A, determine whether A can be satisfied in any of the models. In the presence of completeness with respect to the semantics, the validity and satisfiability problems are dual to each other: each can be reduced to the complement of the other:

$$\begin{aligned} \mathsf{L} \vdash A &\iff \text{NOT } (\neg A \text{ is satisfiable}); \\ A \text{ is satisfiable} &\iff \text{NOT } (\mathsf{L} \vdash \neg A). \end{aligned}$$

The immediate consequence is that decidability of the validity problem is equivalent to the decidability of the satisfiability problem, which is why we usually talk about the decidability of a logic rather than about the decidability of one of these problems for the logic. However, complexity of the two problems may differ. We start this chapter by recalling the way the complexity is measured.

It is common to measure complexity based on some (usually unspecified) type of Turing machines. The exact details of these machines, e.g., the number of tapes, the transition function, etc., are generally irrelevant, with one important exception. It plays a role whether the machine is *deterministic*, i.e., has at most one action possible in each state, or not, i.e., may have more than one possible action. In the latter case, Turing machines are further distinguished by the acceptance conditions: which conditions on the multiple possible runs on a particular input signify the acceptance of this input. However, for the purposes of this book, it is sufficient to consider only one type of Turing machines that are not deterministic, the *nondeterministic* Turing machines. They accept an input if at least one possible run on this input accepts.

Lemma 5.1. *If a problem belongs to a complexity class based on deterministic Turing machines, then its complement belongs to the same class.*

Proof. It is sufficient to flip the acceptance conditions. □

This lemma shows that the only cases when validity and satisfiability problems may belong to different complexity classes is when these classes are based on Turing machines that are not deterministic, e.g., on nondeterministic ones. For instance, if a nondeterministic machine accepts because of some accepting run but other rejecting runs exist, then flipping the acceptance conditions for each run does not lead to a rejection overall.

The description of algorithms is almost never done in Turing machine code. Instead the Church–Turing these is invoked to claim that a detailed enough description of any algorithm in pseudocode can always be transformed into an appropriate Turing machine. We follow the same route.

The complexity is typically measured in one of two ways:

- how much running time is sufficient for (a Turing machine of an appropriate type) to solve the problem;

- how much (tape) space is sufficient for (a Turing machine of an appropriate type) to solve the problem.

In the former case, the decision algorithm has to terminate on all inputs within the allotted time. In the latter case, the time is not bounded, but

5.1 Short Introduction to Complexity Classes

the termination on all inputs is still required. For the complexity of validity/satisfiability problems, the bounds on time/space are formulated in terms of a class of functions rather than specific functions. For all the logics we discuss these bounds will be polynomial. There is a certain sporting interest in determining the lowest polynomial degree sufficient, but we do not engage in such endeavors, partly because for most logics, including the classical propositional logic, the (worst-case) complexity is infeasible from the point of view of implementation anyway. Thus, it is sufficient to paint this infeasibility in wide strokes.

5.1.1 Class P: Deterministic Polynomial Problems

This class consists of all problems that can be solved by a deterministic Turing machine in polynomial time, i.e., there exists a polynomial $p(x)$ such that on any input of size n the machine terminates no later than after $p(n)$ steps. An example such a problem is whether two schemes of formulas X and Y can be unified. Another example is sorting a list. Unfortunately, most logics are too complex to be decided in polynomial time. Even the classical propositional logic is not known to belong to P.

This is the basic or zero class of the complexity hierarchy usually used for validity/satisfiability problems. In other words, this hierarchy is defined modulo deterministic polynomial-time transitions and, hence, is called the *polynomial hierarchy*. There are multiple other complexity hierarchies, both finer- and coarser-grained than this but they remain outside the scope of this book.

Definition 5.2 (Polynomial-time many-one reduction). We say that a decision problem A in the language $\mathcal{L}_A$ *polynomial-time many-one reduces*, or *p-reduces*, to a decision problem B in the language $\mathcal{L}_B$, denoted $A \leq_m^p B$, if there is a function $f \colon \mathcal{L}_A \to \mathcal{L}_B$ computable by a deterministic polynomial-time Turing machine such that for any $w \in \mathcal{L}_A$

$$w \in A \iff f(w) \in B \ .$$

In this case, it is understood that the problem B is *at least as hard as* the problem A because, were B easier than A, this easier decision algorithm for B could be used to decide A by employing an additional polynomial function f, i.e., without increase in complexity.

Definition 5.3 (Hardness, completeness). For a complexity class CP (from the polynomial hierarchy), a problem A is called *CP-hard* if each problem in the class CP p-reduces to A. The problem A is called *CP-complete* if it is CP-hard and itself belongs to the class CP.

In other words, a problem is called CP-complete if it is one of the hardest problems in the class CP, i.e., if there are no strictly harder problems in the class.

Lemma 5.4. *Let A be an arbitrary decision problem from the class P and B an arbitrary non-trivial[1] decision problem. Then $A \leq_m^p B$.*

Proof. Since B is non-trivial, there are $w_1 \in B$ and $w_2 \in \overline{B}$. The polynomial reduction from A to B is defined as follows:

$$f(u) := \begin{cases} w_1 & \text{if } u \in A, \\ w_2 & \text{otherwise.} \end{cases}$$

The trick is that the condition $u \in A$ can be determined in polynomial time by the decision algorithm for A. $\square$

Corollary 5.5. *Any non-trivial decision problem is P-hard. Any non-trivial decision problem from the class P is P-complete.*

This corollary formalizes the intuition that the class P serves as a 0 with respect to complexity both in the sense of minimality and in the sense of being a neutral element.

Lemma 5.6 (Transitivity of p-reducibility). *If $A \leq_m^p B$ and $B \leq_m^p C$, then $A \leq_m^p C$.*

Proof. Given the p-reductions f from A to B and g from B to C, it is easy to see that $g \circ f$ yields a p-reduction from A to C. $\square$

5.1.2 Class NP: Nondeterministic Polynomial Problems

This class NP consists of all problems that can be solved by a nondeterministic Turing machine in polynomial time. In other words, there exists a polynomial $p(x)$ such that all runs on any input of size n must terminate no later than after $p(n)$ steps. Since a deterministic Turing machine can be seen as a nondeterministic one with just one possible run, it is clear that $P \subseteq NP$. Whether $P = NP$ or $P \subsetneq NP$ is one of the most difficult problems facing computer science. It is one of the Clay Institute Millennium problems.

A classical result of complexity theory is the following:

---
[1] A problem is non-trivial if neither it nor its complement is empty.

5.1 Short Introduction to Complexity Classes

Theorem 5.7 (Cook [44], Levin [104]). *The satisfiability problem for classical propositional logic* CL *is NP-complete. This problem is commonly referred to as SAT.*

Cook–Levin theorem claims two things: that SAT can be decided by a nondeterministic Turing machine in polynomial time and that any other problem from NP can be reduced to SAT. The latter statement is proved by a tour de force encoding of an arbitrary nondeterministic Turing machine with given polynomial bounds on termination and its acceptance as a propositional formula. It is an insightful piece of work but would not be useful in this book. Thus, we omit it and direct an interested reader to the original paper by Cook [44].

The former statement, on the contrary, will have to be reproved for more complex logics. Thus, we describe a simple nondeterministic algorithm behind it that is suitable for generalization to these more complex logics.

Two proofs that SAT is in NP. The goal is to find a nondeterministic algorithm that, given a propositional formula A, accepts in time bounded by a polynomial of $|A|$ iff A is satisfiable for some assignment of truth values to atomic propositions. Perhaps, the simplest available algorithm makes nondeterministic choices for truth values of all atomic propositions occurring in A, computes the truth value of A, and accepts if A is true (for at least one set of choices). In the common complexity theory lingo, one says that the algorithm nondeterministically *guesses* truth values of atomic propositions that make A true and then verifies that A is true in polynomial time. □

However, this algorithm is not optimal for applications to LP_{CS} because it is too CL-centric. We now describe a better match for our purposes, an algorithm that is essentially a tableau procedure and, as such, allows for generalizations to modal and justification logics. The algorithm starts with a single-node tableau with TA in the node. The algorithm builds one branch of a tableau proof keeping track of the status of each tableau node on this branch, which can be either **processed** or **unprocessed**. (This branch can be represented by a list of signed formulas with a bit attribute for the processing status.) The initial node with TA is in the state **unprocessed**. While there are **unprocessed** nodes, the algorithm nondeterministically guesses one of the unprocessed nodes, changes its status to **processed**, and tries to apply a tableau construction rule to it (see Definition 1.14). More precisely,

- if the node contains $TC \to D$ and neither FC nor TD is present on the branch so far, then the algorithm nondeterministically chooses whether to add FC or TD (with state **unprocessed**) to the branch;

- if the node contains $FC \to D$ and at least one of TC or FD is absent from the branch so far, then the algorithm adds (with state **unprocessed**) all elements of the set $\{TC, FD\}$ that are missing from the branch.

After all formulas on the branch have status **processed**, the algorithm checks whether the constructed branch is open. The algorithm accepts if this (at least one so constructed) branch is open. Since only subformulas of A can occur on the branch and each formula is processed at most once, it follows that there are at most $|A|$ steps of the algorithm. It is clear that each step requires at most quadratic time (note that the applicability check generally requires comparing a formula of size at most $|A|$ with all the formulas on the branch, whose length is at most $|A|$). It is also clear that a trivial check whether the branch is closed can be done in quadratic time in the length of the branch, i.e., in $O(|A|^2)$. □

Corollary 5.8. *If a logic* L *is conservative over* CL, *then the satisfiability problem for it is NP-hard.*

Proof. Any problem in NP can be p-reduced to SAT because it is NP-hard. Due to conservativity, the identity function provides a polynomial-time reduction from SAT to the satisfiability for L. Thus, by Lemma 5.6, any function p-reducing a problem from NP to SAT also p-reduces the same problem to the satisfiability for L. □

Hence, by Theorems 1.29 and 2.13

Corollary 5.9. *The satisfiability problem for both* S4 *and* LP_{CS} *is NP-hard.*

Such statements of hardness are regarded as lower complexity bounds stating that determining satisfiability in for S4 and for LP_{CS} is at least as hard as any problem in NP but may be harder. Statements of completeness such as the Cook–Levin Theorem 5.7 consist of a lower complexity bound (SAT is NP-hard) and an upper complexity bound (SAT is in NP). Thus, completeness statements provide optimal complexity bounds. While it is not known whether the lower complexity bounds for S4 and for LP_{CS} are optimal, we will soon see that there is a reason to believe they are not.

5.1.3 Class coNP: Dual to NP

Since NP is based on nondeterministic Turing machines, the complements of problems in NP need not belong to NP. The class containing them is called coNP. In particular, if the satisfiability problem for a logic is in NP, then its validity problem is in coNP.

Lemma 5.10. *If $A \leq^p_m B$, then $\overline{A} \leq^p_m \overline{B}$.*

Proof. Any f that p-reduces A to B also p-reduces $\overline{A}$ to $\overline{B}$. □

Corollary 5.11. *A complement of any NP-hard (NP-complete) problem is coNP-hard (coNP-complete) and vice versa.*

Corollary 5.12. *The validity problem for CL is coNP-complete. The validity problem for both S4 and $\mathsf{LP_{CS}}$ is coNP-hard.*

Given the definition of NP as the class of problems that can be solved by polynomial-time nondeterministic Turing machines that *accept* iff at least one of their runs *accepts*, it is easy to see that coNP consists of problems that can be solved by polynomial-time nondeterministic Turing machines that *reject* iff at least one of their runs *rejects*. In particular P ⊆ coNP. Since we don't know whether P is strictly smaller than NP, we cannot know whether P is strictly smaller than coNP. There is, however, an additional independent uncertainty regarding the relation of NP to coNP: it is not known whether NP = coNP. Obviously, this would be true if P = NP. But it could be true otherwise too.

5.1.4 Class PSPACE: Polynomial-Space Problems

The class PSPACE consists of all problems that can be solved by a deterministic Turing machine using polynomial space, i.e., there must exist a polynomial $p(x)$ such that on each input of size n the machine terminates after using no more than $p(n)$ tape cells.

Since it is impossible to use more than polynomially many cells in polynomial time, P ⊆ PSPACE. As usual, it is not known whether the inclusion is strict. For the same reason, it should be clear that NP ⊆ NPSPACE, where NPSPACE is obtained from PSPACE by replacing deterministic machines with nondeterministic ones. However, using the following non-trivial statement:

Theorem 5.13 (Savitch [123]). *NPSPACE = PSPACE.*

Corollary 5.14. $NP \subseteq PSPACE$ and $coNP \subseteq PSPACE$.

Proof. The first statement is immediate by Savitch's theorem. The second statement now follows by Lemma 5.1. □

For both inclusions, it is not known whether they are strict.

Theorem 5.15 (Ladner [102]). *Both the validity and the satisfiability problems for* S4 *is PSPACE-complete.*

Given the Gödel translation from IPC to S4, it should come as no surprise that

Theorem 5.16 (Statman [128]). *Both the validity and the satisfiability problems for* IPC *is PSPACE-complete.*

However, we need one more class to describe the complexity of LP_{CS}.

5.1.5 Oracle Computations and Polynomial Hierarchy

An oracle Turing machine has an additional capacity for performing 1-step oracle queries regarding some fixed decision problem.

Definition 5.17 (Classes Σ_2^p and Π_2^p). The class Σ_2^p consists of all problems that can be solved by an NP oracle machine with queries regarding some fixed problem from NP, i.e., by a nondeterministic oracle machine with polynomial running time and with queries to an NP problem. The class Π_2^p consists of complements of problems from Σ_2^p.

Similar to the independent description of coNP, we can describe Π_2^p as consisting of all problems that can be solved by a polynomial-time Turing machine that can use an oracle for a problem from NP and that rejects iff at least one run rejects, i.e., solved by a coNP oracle machine with queries to an NP problem.

Moreover, it clearly does not matter whether queries are to a problem A or to its complement. Indeed, to obtain answers to A-queries, it is sufficient to invert the answers to $\overline{A}$-queries. If we denote by Φ^Ψ the class of problem solvable by Φ oracle machines with queries to a fixed problem from Ψ for given complexity classes Φ and Ψ, we get the following properties:

$$\Sigma_2^p = NP^{NP} = NP^{coNP} \; , \quad \Pi_2^p = coNP^{NP} = coNP^{coNP} \; ; \quad (5.1)$$

$$NP \cup coNP \subseteq P^{NP} = P^{coNP} \; ; \quad (5.2)$$

$$NP^P = NP \; , \quad coNP^P = coNP \; . \quad (5.3)$$

5.2 Complexity of LP$_{CS}$

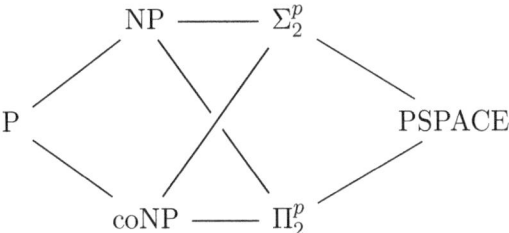

Figure 5.1: Relationships among several complexity classes: each line represents known non-strict inclusions from left to right

Lemma 5.18. $NP \subseteq \Sigma_2^p$ and $coNP \subseteq \Pi_2^p$.

Proof. Trivial since the oracle need not be used. □

Lemma 5.19. $NP \subseteq \Pi_2^p$ and $coNP \subseteq \Sigma_2^p$.

Proof. Let A be a problem from NP. Then it can be used as an oracle in Π_2^p. Thus, in a Π_2^p program it is sufficient to query that oracle.
Let B be a problem from coNP. By (5.1) it can be used as an oracle in Σ_2^p, making the situation analogous to the one just discussed. □

Once again, for none of these inclusions is it known whether it is strict though most papers are written under the implicit assumption that they are. The relationships among these complexity classes is depicted in Figure 5.1. The subscripts 2 in Σ_2^p and Π_2^p suggest that the process of supplying NP oracles can be iterated to produce a countable hierarchy of classes all of which are contained in PSPACE. This hierarchy is called the *polynomial hierarchy* and explains the superscript p in the names of classes. In terms of this hierarchy NP is Σ_1^p and coNP is Π_1^p, i.e., these two classes belong to the first level of the hierarchy. We omit the definitions and the discussion of the levels higher than the second one.

5.2 Complexity of LP$_{CS}$

As we can see from Figure 5.1, the validity/satisfiability problems for intuitionistic propositional logic IPC and the modal logic S4 are (believed to be) harder than those for classical propositional logic CL. It is our goal to find the place of the logics LP$_{CS}$ in this picture. We know from Corollary 5.12 that the validity problem must be at least coNP-hard (and the satisfiability problem at least NP-hard). It should be clear from Theorem 4.37 that the complexity of $\mathsf{LP_{CS}}$ depends on the constant specification CS. Fortunately, it is possible to relegate this dependency to an oracle, which, in effect, determines the complexity of the logic.

5.2.1 Master Algorithm

For the classical propositional logic we have described a tableau-based NP algorithm for SAT in Sect. 5.1.2. We now show how to modify this algorithm and supply it with an oracle obtain an oracle-based "algorithm" for any $\mathsf{LP_{CS}}$.[2] First, we need to add tableau construction rules for formulas $t{:}C$.

Algorithm 5.20 (Satisfiability of $\mathsf{LP_{CS}}$). The branch-extending steps of the algorithm are as follows: the algorithm nondeterministically chooses one of the nodes with status unprocessed, changes its status to processed and

- if the node contains $TC \to D$ and neither FC nor TD is present on the branch so far, then the algorithm nondeterministically chooses whether to add FC or TD to the branch;

- if the node contains $FC \to D$ and at least one of TC or FD is absent from the branch so far, then the algorithm adds all elements of the set $\{TC, FD\}$ that are missing from the branch;

- if the node contains $T\,t{:}C$ and TC is not present on the branch so far, then the algorithm adds TC to the branch;

- if the node contains $F\,t{:}C$ and FC is absent from the branch so far, then the algorithm non-deterministically determines whether to make this processed $F\,t{:}C$ active or to add unprocessed FC to the branch.

After all formulas on the branch have status processed, the algorithm submits the branch to the oracle with a query. *Are both of the following conditions fulfilled:*

1. *the constructed branch is open;*

---

[2] We put the word algorithm in quotation marks because, in some cases, the oracle is queried regarding an undecidable problem.

5.2 Complexity of LP$_{CS}$

2. for
$$\mathcal{B} := \mathsf{CS} \cup \{(t, C) \mid Tt{:}C \text{ is on the branch}\} \qquad (5.4)$$

and $\mathcal{E}(\mathcal{B})$ being the least fixed point of the operator $\mathsf{cl}_\mathcal{B}$ from Definition 4.4, which exists by Lemma 4.5,

$$(s, D) \notin \mathcal{E}(\mathcal{B}) \text{ for any \textbf{active} } Fs{:}D \text{ on the branch}. \qquad (5.5)$$

The algorithm accepts if the oracle answers "yes" (for at least one so constructed branch).

We now show the correctness of this NP algorithm with oracle for the satisfiability problem:

Lemma 5.21. *The above-stated algorithm accepts iff the given formula is satisfiable in* LP$_{CS}$.

Proof. Assume first that both conditions of the oracle query are satisfied. Define $\mathsf{val} := \{P \mid TP \text{ is on the branch}\}$ and $\mathcal{M} := (\mathsf{val}, \mathcal{B})$ for $\mathcal{B}$ from (5.4). Clearly, $\mathcal{M}$ is a finitary CS-model. It is easy to show

$$\left.\begin{array}{l}\mathcal{M} \Vdash E \text{ for every } TE \\ \mathcal{M} \nVdash E \text{ for every } FE\end{array}\right\} \text{ on the branch} \qquad (5.6)$$

by simultaneous induction on the size of E. In particular, it would follow that $\mathcal{M} \Vdash A$ since the all branches begin with TA. We only show the case of $E = t{:}C$. If $Tt{:}C$ is on the branch, so is TC, making $\mathcal{M} \Vdash C$ by the induction hypothesis. Further, $(t, C) \in \mathcal{B} \subseteq \mathcal{E}(\mathcal{B})$. It follows that $\mathcal{M} \Vdash t{:}C$. If an **active** $Ft{:}C$ is on the branch, then $(t, C) \notin \mathcal{E}(\mathcal{B})$ by (5.5), and $\mathcal{M} \nVdash t{:}C$. If $Ft{:}C$ is on the branch but is not **active**, then FC is also on the branch, and again $\mathcal{M} \nVdash t{:}C$.

It remains to show the direction from right to left: if a formula is satisfiable, then there is a branch, satisfying both conditions. Assume that A is satisfiable. By Theorem 4.35, there exists a finitary CS-model $\mathcal{M} = (\mathsf{val}, \mathcal{B}')$ such that $\mathcal{M} \Vdash A$. Thus, (5.6) is fulfilled for the initial branch TA. It is easy to prove that after each step of the algorithm, there is a branch such that (5.6) remains satisfied for $\mathcal{M}$. Again, we only consider the case of processing $E = t{:}C$ on a branch satisfying this condition and leave other cases as an easy exercise. When $Tt{:}C$ from the branch satisfying this condition is processed, TC is added to the branch. By the induction hypothesis, $\mathcal{M} \Vdash t{:}C$, in particular, $\mathcal{M} \Vdash C$, making (5.6) true for the extended branch. When $Ft{:}C$ is processed, there are two possibilities: $\mathcal{M} \not\Vdash C$ or $\mathcal{M} \Vdash C$. In the former case, choose the branch extended with FC where $Ft{:}C$ is not active; in the latter case, choose the branch that makes $Ft{:}C$ active. Clearly, (5.6) remains true in either case.

After all formulas are processed, we need to show that both conditions for the acceptance are fulfilled for the branch satisfying (5.6). Indeed, the openness is guaranteed because having TB and FB on a branch would imply that $\mathcal{M} \Vdash B$ and $\mathcal{M} \not\Vdash B$, which is impossible. Further, for $\mathcal{B}$ from (5.4), it is clear that $\mathcal{B} \subseteq \mathcal{E}(\mathcal{B}')$. Indeed, $\mathsf{CS} \subseteq \mathcal{B}' \subseteq \mathcal{E}(\mathcal{B}')$ because $\mathcal{M}$ is a finitary CS-model. For each $(t, C) \in \mathcal{B}$ because of the presence of $Tt{:}C$ on the branch, we know that $\mathcal{M} \Vdash t{:}C$ and, in particular, $(t, C) \in \mathcal{E}(\mathcal{B}')$. By monotonicity and idempotence of $\mathcal{E}$ (Lemmas 4.7 and 4.8), it follows that $\mathcal{E}(\mathcal{B}) \subseteq \mathcal{E}(\mathcal{B}')$. For every $Fs{:}D$ on the branch, we have $\mathcal{M} \not\Vdash s{:}D$, creating two possibilities: either $\mathcal{M} \not\Vdash D$ or $(s, D) \notin \mathcal{E}(\mathcal{B}')$. In the latter case, $(s, D) \notin \mathcal{E}(\mathcal{B})$ is immediate. In the former case, $Fs{:}D$ was not made active, so (5.5) is vacuously true. □

Lemma 5.22. *Algorithm 5.20 is an NP algorithm with oracle.*

Proof. Since oracle computation is treated as a black box, all we need to show is that the construction of each branch is polynomial. As in the purely propositional case (see the second proof of Theorem 5.7), only subformulas of the given formula A can occur on the branch and each formula is processed at most once, it follows that there are at most $|A|$ steps of the algorithm. It is clear that the two new types of steps also require at most quadratic time for the applicability check and at most linear time for the implementation. □

The proof of the Master Theorem shows that the decision algorithm for the satisfiability of $\mathsf{LP_{CS}}$ can be organized as a succession of a(n essentially) propositional satisfiability check followed by an additional verification of consistency for justification-related elements. This both provides an alternative explanation for the general lower complexity bounds for all CS and pinpoints the culprit for the increase in complexity observed in justification logics for standard classes of CS.

5.2 Complexity of LP$_{\text{CS}}$

5.2.2 Best Case: Finite CS

This is the case when the complexity of LP$_{\text{CS}}$ is the same as that of CL. Because of (5.3), to demonstrate this it is sufficient to show that the oracle queries can be solved by a deterministic polynomial-time algorithm.

Lemma 5.23. *Let CS be a finite constant specification. There exists a P algorithm[3] that can, for any branch constructed by Algorithm 5.20 on input A, check whether the branch is open and, at the same time, (5.5) is satisfied.*

Proof. Checking whether the branch is open can clearly be done in time quadratic in the size of the branch. It remains to explain how to check the newly-established acceptance condition (5.5) in deterministic polynomial time. We discussed in some detail how to decide (5.5) for each individual (s, D) in Lemma 4.30 for the much more general case of almost schematic CS. It remains to show how to do this efficiently, i.e., in polynomial time for all active $Fs{:}D$ on the branch. Since the number of active $Fs{:}D$ on the branch is bounded by the number of subformulas of A, i.e., is linear in $|A|$, it is sufficient to show that (5.5) can be determined in polynomial time for any one active $Fs{:}D$ on the branch. In this proof, we reuse the notation from the proof of Lemma 4.30.

First of all, since our finite constant specification consists of the empty schematic part $\text{CS}_1 = \varnothing$ and finite part $\text{CS}_2 = \text{CS}$, we get

$$\text{CS}^s = \varnothing^s \cup \text{CS} = \text{CS} \ .$$

In other words, there is no need to extend the language with schematic variables over formulas. Everything can be done on the level of formulas. Note also that CS is of fixed finite size, i.e., $O(1)$ in $|A|$.

Further, by (4.7), it is sufficient to construct $\|s\|^{\text{rk}(s)}$ and check whether D is represented there. Let us first show that $\|s\|^{\text{rk}(s)}$ can be constructed in polynomial time. Since $\text{rk}(s)$ is also linear in $|A|$, it is sufficient to show that, for each $0 \leq i \leq \text{rk}(s)$, we can simultaneously construct in polynomial time $\|s'\|^i$ for all subterms s' of s. Note that the number of these subterms is also linear in $|A|$. Going through the proof of Lemma 4.30, it is easy to see that the union of all $\|s'\|^0$ is polynomial in $|A|$ (finitely many formulas from CS and polynomially many

---

[3] While the input of this algorithm is the whole branch as opposed to the input A to Algorithm 5.20, the branch is constructed in polynomial time and, hence, has size polynomial in $|A|$. Thus, any algorithm requiring polynomial time in the size of the branch also works time polynomial in $|A|$.

from the given branch) and consists of pure formulas, without the use of schematic variables. Thus, computing them all requires polynomial time. Each step from i to $i+1$ requires linearly many computations of $\|s'\|^{i+1}$. Thus, it is sufficient to show that for each s' we can perform the step in polynomial time. As can be seen from Def. 4.25, the cases of $s' = !s''$ and $s' = s'' + s'''$ are reasonably simple and are left for the reader. Note that they do not introduce schematic variables. Note also that the size of formulas in $\|s'\|^{i+1}$ is not increased compared to those in $\|s''\|^i$ and $\|s'''\|^i$ at all for the case of $+$ and is increased by at most $|A|$ in the case of !. Consider the case of $s' = s'' \cdot s'''$. In the absence of schematic variables on step i, Case Def. 4.25.3a does not occur and the unification of the schemes U_1 with U_2 in Case Def. 4.25.3b degenerates to equality of formulas. In other words, we simply set

$$\|s'' \cdot s'''\|^{i+1} := \|s'' \cdot s'''\|^i \cup \{G \mid F \to G \in \|s''\|^i \text{ and } F \in \|s'''\|^i\} \ .$$

This operation is quadratic in the joint sizes of $\|s''\|^i$ and $\|s'''\|^i$ and the resulting formulas G are smaller than $F \to G$ present at the preceding step.

Finally, since $\|s\|^{\mathsf{rk}(s)}$ consists of polynomially many pure formulas, checking whether D is represented there is a linear operation in its size or a polynomial one in $|A|$. □

Corollary 5.24. *The satisfiability problem for* $\mathsf{LP}_{\mathsf{CS}}$ *with a finite* CS *is NP-complete, and the validity problem is coNP-complete.*

Proof. By Lemmas 5.21, 5.22, and 5.23, the satisfiability problem is in NP^P, which is the same as NP by (5.3). By Corollary 5.9, it is also NP-complete. Hence, by Corollary 5.11, the validity problem is coNP-complete. □

5.2.3 Worst Case: Irregular CS

We have already established in Theorem 4.37 that $\mathsf{LP}_{\mathsf{CS}}$ can be undecidable. The existence of an NP algorithm with an oracle for the undecidable problem might seem counter-intuitive. But there is no mystery here. It simply means that the problem stated in (5.5), which the oracle has to solve, is undecidable for the constant specification from the proof of Theorem 4.37.

5.2.4 Standard Case: Almost Schematic CS

We show the upper complexity bound.

Lemma 5.25. *Let CS be a decidable and almost schematic constant specification. There exists a coNP algorithm that can, for any branch constructed by Algorithm 5.20 on input A, check whether the branch is open and, at the same time, (5.5) is satisfied.*

Corollary 5.26. *The satisfiability problem for $\mathsf{LP_{CS}}$ with a decidable and almost schematic CS is in Σ_2^p, and the validity problem is in Π_2^p.*

Before proving Lemma 5.25, it is instructive to understand why the polynomial-time implementation of the algorithm from the proof of Lemma 5.23 cannot be applied to almost schematic CS. The obvious difference is that here we are forced to deal with schematic variables over formulas and unification, just like in Lemma 4.30. But why can't the polynomial handling of formulas be extended to schemes? The answer may not be so trivial. One might blame the size of formulas assigned to terms $s'' \cdot s'''$. We noted in the proof of Lemma 5.23 that the size of formulas G added to $\|s'' \cdot s'''\|^{i+1}$ is strictly less that that of formulas $F \to G$ already assigned to s''. This nice property fails in the presence of schematic variables over formulas. Here is a simple example to illustrate this phenomenon: let U be a schematic variable over formulas and V an arbitrary scheme and

$$U \to \underbrace{U \vee \cdots \vee U}_{k} \in \|s''\|^i \quad \text{and} \quad \underbrace{V \wedge \cdots \wedge V}_{l} \in \|s'''\|^i .$$

After the unification of U with $\underbrace{V \wedge \cdots \wedge V}_{l}$, we must assign

$$\underbrace{(\underbrace{V \wedge \cdots \wedge V}_{l}) \vee \cdots \vee (\underbrace{V \wedge \cdots \wedge V}_{l})}_{k} \in \|s'' \cdot s'''\|^{i+1} .$$

In other words, two schemes of sizes $O(k)$ and $O(l)$ produce a scheme of size $O(kl)$ after a single step. Even if all initial schemes at step $i = 0$ are taken from axioms and, hence, have size no greater than some fixed number K, this seems to spell trouble. Indeed, after $\mathsf{rk}(s)$ steps, the scheme size can grow as much as $K^{\mathsf{rk}(s)}$, which is generally exponential in the size of A.

This problem is, however, not new and have been solved a long time ago (see, e.g., [45]). The trick is to use a more efficient data structure for storing schemes. Instead of syntactic trees that grow exponentially primarily due to multiple repetitions, one can obtain a polynomial algorithm through the use of syntactic dags (direct acyclic graphs). In other words, it is allowed to store two occurrences of the same subformula in the same storage space.

The real culprit is not the size but the number of formulas and schemes assigned to each term. In the case of finite CS it is not hard to show that the number of formulas assigned to each term on each step remains linear in $|A|$. Coupled with a polynomial-size representation of each formula, linear number of terms, and linear number of steps, this requires polynomial storage for all the formulas assigned to all the terms during all the steps.

And the crucial difference is in the processing of $\cdot$. In the case of finite CS, the number of formulas added to $s'' \cdot s'''$ is no larger than the number of formulas assigned to s'' on the previous step: each formula $F \to G$ in $\|s''\|^i$ may produce at most one formula in $\|s'' \cdot s'''\|^{i+1}$, namely G. On the other hand if

$$\|s''\|^i = \{U_1 \to W_1, \ldots, U_k \to W_k\} \quad \text{and} \quad \|s'''\|^i = \{V_1, \ldots V_l\},$$

then potentially up to kl new schemes may be added to $\|s'' \cdot s'''\|^{i+1}$: for each pair of i and j such that U_i can be unified with V_j, the algorithm should add $W_i \sigma_{ij}$, where σ_{ij} is the mgu of U_i and V_j. In other words, starting with at most K schemes assigned to each term, the algorithm may have to deal with $K^{\mathsf{rk}(s)}$ schemes assigned to s, an exponential blow-up.

Unable to treat exponentially many disparate schemes in a polynomial manner, the algorithm resorts to checking each combination in turn, resulting in a coNP dynamics. In other words instead of constructing all formulas assigned to a subterm, the algorithm will be considering at most one scheme assigned to it per run.

Proof of Lemma 5.25. The openness check is polynomial as in the case of a finite constant specification. Since a coNP algorithm for (5.5) is easily obtained from an NP algorithm for the problem's complement, it is sufficient to describe an NP algorithm for determining that the desired algorithm should return failure. Nondeterministically choose one of polynomially-many active $Fs{:}D$ on the branch. For each of linearly

5.2 Complexity of LP$_{CS}$

many nodes σ in the syntactic tree $\mathfrak{S}$ of s, nondeterministically assign one of two possible attributes: branch or CS. For each of linearly many branch nodes $\sigma \in \mathfrak{S}$, nondeterministically choose one of polynomially many formulas C such that $Tt_\sigma{:}C$ is on the branch if such a C exists. If so, add (σ, C) to the set EB being constructed by the algorithm. For each of linearly many CS nodes $\sigma \in \mathfrak{S}$, nondeterministically choose one of finitely[4] many schemes U such that $(t_\sigma, U) \in \mathsf{CS}^s$ if such a U exists. If so, add (σ, U) to the set EB being constructed by the algorithm. For each leaf σ of $\mathfrak{S}$ that still has no entry $(\sigma, V) \in EB$ add $(\sigma, \varnothing)$ to EB.

We say that EB *covers* a node $\sigma \in \mathfrak{S}$ if some pair with σ as the first element belongs to EB. We say that EB *contains no conflicts* if no two pairs with the same first element exist in it.

After this initialization stage, EB covers each leaf of $\mathfrak{S}$ and contains no conflicts. While EB does not cover all nodes of $\mathfrak{S}$, the algorithm performs the following actions, which can easily be shown to introduce no new conflicts:

1. Nondeterministically choose[5] one of linearly many $\sigma \in \mathfrak{S}$ such that EB does not yet cover σ but covers all of its children in $\mathfrak{S}$.

2. If $+$ is the primary connective of t_σ, then $t_\sigma = t_{\sigma_0} + t_{\sigma_1}$ for the only two children σ_0 and σ_1 of σ in $\mathfrak{S}$.

   - If $\{(t_{\sigma_0}, U_0), (t_{\sigma_1}, U_1)\} \subseteq EB$ for some schemes U_0 and U_1, then nondeterministically choose i to be 0 or 1 and add the pair (t_σ, U_i) to EB.
   - If $\{(t_{\sigma_i}, U), (t_{\sigma_{1-i}}, \varnothing)\} \subseteq EB$ for some $i = 0, 1$ and some scheme U, then add (t_σ, U) to EB.
   - Otherwise, i.e., in case $\{(t_{\sigma_0}, \varnothing), (t_{\sigma_1}, \varnothing)\} \subseteq EB$, add $(t_\sigma, \varnothing)$ to EB.

3. If $\cdot$ is the primary connective of t_σ, then $t_\sigma = t_{\sigma_0} \cdot t_{\sigma_1}$ for the only two children σ_0 and σ_1 of σ in $\mathfrak{S}$.

   - If $\{(t_{\sigma_0}, \Upsilon_j^\kappa), (t_{\sigma_1}, U_1)\} \subseteq EB$ for some descendant κ of σ_0 in $\mathfrak{S}$, some schematic variable Υ_j^κ over formulas and some scheme U_1, add $(t_\sigma, \Upsilon_0^\sigma)$ to EB for the schematic variable Υ_0^σ over formulas.

---

[4]More precisely, finite here means $O(1)$ of the input size.
[5]This nondeterminism is nonessential: it can be avoided by defining a particular processing order from leaves to the root.

- If $\{(t_{\sigma_0}, V \to W), (t_{\sigma_1}, U_1)\} \subseteq EB$ for some schemes U_1, V, and W, then find the mgu τ of V and U_1 using our fixed unification algorithm and add $(t_\sigma, W\tau)$ to EB if such an mgu exists or add $(t_\sigma, \varnothing)$ otherwise.

- Otherwise, i.e., if

$$(t_{\sigma_0}, \varnothing) \in EB \quad \text{or} \quad (t_{\sigma_1}, \varnothing) \in EB \quad \text{or} \quad (t_{\sigma_0}, U) \in EB$$

but U is neither a schematic variable over formulas nor an implication, add $(t_\sigma, \varnothing)$ to EB.

4. If ! is the primary connective of t_σ, then $t_\sigma = !t_{\sigma_1}$ for the single child σ_1 of σ in $\mathfrak{S}$.

  - If $(t_{\sigma_1}, U) \in EB$ for some scheme U, then add $(t_\sigma, t_{\sigma_1}{:}U)$ to EB.

  - Otherwise, i.e., if $(t_{\sigma_1}, \varnothing) \in EB$, add $(t_\sigma, \varnothing)$ to EB.

It is not hard to show that this while loop always terminates resulting in EB that covers every node of $\mathfrak{S}$. In particular, EB covers the root ρ of $\mathfrak{S}$, which corresponds to $t_\rho = s$.

The algorithm for determining (5.5) returns **failure** if, for some **active** $Fs{:}D$ on the tableau branch, $(\rho, U) \in EB$ for some scheme U that unifies with D (for one choice of D and the consequent set of nondeterministic choices). Given that unification can be done in polynomial time (both finding an mgu and applying a substitution to a scheme), it is not hard to show that each branch of the algorithm requires only polynomial time. It remains to show that

the algorithm returns failure iff (5.5) is violated.

By (4.6), we can rewrite the second part of the equivalence as

D is an instance of some $U \in \|s\|^i$

for some **active** $Fs{:}D$ on the tableau branch

and some non-negative i. (5.7)

Since the unification of D with the Us constructed by the algorithm is performed explicitly, it is sufficient to show that, for each **active** $Fs{:}D$ on the tableau branch,

$$(\exists i)U \in \|s\|^i \iff (\rho, U) \in EB \text{ for some branch of the algorithm.} \tag{5.8}$$

We start with the easier direction, from right to left. To demonstrate it, it is sufficient to prove the following statement by induction on the depth of σ in $\mathfrak{S}$. We prove that for any scheme V

$(\sigma, V) \in EB$ for some branch of the algorithm

implies $(\exists i) V \in \|t_\sigma\|^i$.

- If (σ, C) is added to EB because $Tt_\sigma{:}C$ is on the tableau branch, then, $(t_\sigma, C) \in \mathcal{B}$ and $C \in \|t_\sigma\|^0$.

- If (σ, V) is added to EB because $(t_\sigma, V) \in \mathsf{CS}^s$, then $(t_\sigma, V) \in \mathcal{B}$ and $V \in \|t_\sigma\|^0$.

- If $(\sigma, t_{\sigma_1}{:}V')$ is added to EB because $t_\sigma = !t_{\sigma_1}$ and $(\sigma_1, V') \in EB$, then, by the induction hypothesis, $V' \in \|t_{\sigma_1}\|^i$ for some i. It follows that $t_{\sigma_1}{:}V' \in \|!t_{\sigma_1}\|^{i+1}$.

- If $(\sigma, W\tau)$ was added to EB because $t_\sigma = t_{\sigma_0} \cdot t_{\sigma_1}$, and

$$(\sigma_0, W' \to W) \in EB ,$$

and $(\sigma_1, V') \in EB$, and τ is the mgu of W' and V' constructed by our fixed unification algorithm, then by the induction hypothesis, $W' \to W \in \|t_{\sigma_0}\|^i$ and $V' \in \|t_{\sigma_1}\|^j$ for some i and j. Let $M = \max(i, j)$. It follows that $W' \to W \in \|t_{\sigma_0}\|^M$ and $V' \in \|t_{\sigma_1}\|^M$ and, hence, $W\tau \in \|t_{\sigma_0} \cdot t_{\sigma_1}\|^{M+1}$.

- If $(\sigma, \Upsilon_0^\sigma)$ for a schematic variable Υ_0^σ over formulas was added to EB because $t_\sigma = t_{\sigma_0} \cdot t_{\sigma_1}$, and because $(\sigma_0, \Upsilon_l^\kappa) \in EB$ for some descendant κ of σ in $\mathfrak{S}$ and some schematic variable Υ_l^κ over formulas, and because $(\sigma_1, V') \in EB$ for some scheme V', then by the induction hypothesis, $\Upsilon_l^\kappa \in \|t_{\sigma_0}\|^i$ and $V' \in \|t_{\sigma_1}\|^j$ for some i and j. Let $M = \max(i,j)$. It follows that $\Upsilon_l^\kappa \in \|t_{\sigma_0}\|^M$ and $V_{\sigma_1} \in \|t_{\sigma_1}\|^M$ and, hence, $\Upsilon_0^\sigma \in \|t_{\sigma_0} \cdot t_{\sigma_1}\|^{M+1}$.

- The case of $t_\sigma = t_{\sigma_0} + t_{\sigma_1}$ is analogous.

Let us now prove the direction from left to right. We prove a slightly stronger statement to ensure independence of choices in the initial stage of the algorithm. We prove that for any node $\sigma \in \mathfrak{S}$ and any scheme $V \in \|t_\sigma\|^i$ for some i there exists a set of choices

- of branch vs. CS for some descendants of σ in $\mathfrak{S}$,

- of formulas taken from the branch for some branch descendants of σ in $\mathfrak{S}$,

- of axiom schemes for some CS descendants of σ in $\mathfrak{S}$, and

- of 0 vs. 1 for some descendants κ of σ such that $t_\kappa = t_{\kappa_0} + t_{\kappa_1}$ for the children κ_0 and κ_1 of κ,

such that each branch of the algorithm making such choices results in $(\sigma, V) \in EB$. We continuously rely on the fact that the algorithm never removes elements from EB. We use induction on i.

- If $V \in \|t_\sigma\|^0$, then $(t_\sigma, V) \in \mathcal{B}$, which may happen for two reasons:

    - $Tt_\sigma{:}V$ is on the tableau branch, where V is an ordinary formula. In this case, it is sufficient to choose branch for σ and choose this V from the tableau branch to ensure that $(\sigma, V) \in EB$;
    - $(t_\sigma, V) \in \mathsf{CS}^s$. In this case, it is sufficient to choose CS for σ and choose this V from CS^s to ensure that $(\sigma, V) \in EB$.

- For any $i > 0$, there are several possibilities why $V \in \|t_\sigma\|^i$:

    - $V \in \|t_\sigma\|^i$ because $V \in \|t_\sigma\|^{i-1}$. In this case, the statement follows directly by the induction hypothesis.
    - $t_\sigma = {!}t_{\sigma_1}$ for the only child σ_1 of σ, $V = t_{\sigma_1}{:}V'$, and
    $$V' \in \|t_{\sigma_1}\|^{i-1}.$$
    By the induction hypothesis, there is a set of choices for some descendants of σ_1 that ensures that $(\sigma_1, V') \in EB$. All descendants of σ_1 are descendants of σ, and the same set of choices also ensures that $(\sigma, t_{\sigma_1}{:}V')$ is added to EB.

5.2 Complexity of LP$_{CS}$

- $t_\sigma = t_{\sigma_0} \cdot t_{\sigma_1}$ for the children σ_0 and σ_1 of σ,

$$W' \to W \in \|t_{\sigma_0}\|^{i-1},$$

$V' \in \|t_{\sigma_1}\|^{i-1}$, and $V = W\tau$ for the mgu τ of W' and V' constructed by our fixed unification algorithm. By the induction hypothesis, there is a set of choices for some descendants of σ_0 that ensures that $(\sigma_0, W' \to W) \in EB$. By the induction hypothesis, there is also a set of choices for some descendants of σ_1 that ensures that $(\sigma_1, V') \in EB$. Since σ_0 and σ_1 are siblings in $\mathfrak{S}$, these choices can be made independently of each other. Hence, the union of these two sets of choices, which only affect descendants of σ, ensures that both $(\sigma_0, W' \to W) \in EB$ and $(\sigma_1, V') \in EB$, which implies that $(\sigma, W\tau) \in EB$ for the mgu τ.

- $t_\sigma = t_{\sigma_0} \cdot t_{\sigma_1}$ for the children σ_0 and σ_1 of σ, $V = \Upsilon_0^\sigma$,

$$\Upsilon_l^\kappa \in \|t_{\sigma_0}\|^{i-1} \text{ for some descendant } \kappa \text{ of } \sigma_0,$$

and $\|t_{\sigma_1}\|^{i-1} \neq \varnothing$. Let V' be one of the elements of $\|t_{\sigma_1}\|^{i-1}$. By the induction hypothesis, there are two independent sets of choices: one, for some descendants of σ_0, that ensures that $(\sigma_0, \Upsilon_l^\kappa) \in EB$ and the other, for some descendants of σ_1, that ensures that $(\sigma_1, V') \in EB$. Hence, the union of these sets of choices, made only for descendants of σ, ensures that $(\sigma, \Upsilon_0^\sigma) \in EB$.

- $t_\sigma = t_{\sigma_0} + t_{\sigma_1}$ for the children σ_0 and σ_1 of σ and $V \in \|t_{\sigma_j}\|^{i-1}$ for some $j = 0, 1$. By the induction hypothesis, there exists a set of choices for descendants of this σ_j, which are also descendants of σ, that ensures that $(\sigma_j, V) \in EB$. Adding to this set the choice of j for σ, a descendant of itself, we can ensure that $(\sigma, V) \in EB$. $\square$

5.2.5 Lower Bounds

In the best and worst case for the complexity, the matching lower bound was already there for the taking. For finite CS, the lower complexity bound of coNP-hardness was inherited from the propositional logic. In the undecidable case, the result was, in effect, a lower bound already. Only the situation with the Π_2^p algorithm for (almost) schematic CS calls for a proof of a non-trivial lower bound.

Milnikel [110] established a lower bound by encoding QBF-2 as a formula in the language of justification logic that is provable exactly if the quantified Boolean formula is true. To give a precise statement of his result, we need the following definition.

Definition 5.27 (Schematically injective CS). A constant specification CS is called *schematically injective* if CS is schematic and each constant justifies no more than one axiom scheme.

We have the following theorem about lower complexity bounds. A proof can be found in [110].

Theorem 5.28. *Let* CS *be a decidable, schematically injective, and axiomatically appropriate constant specification. The satisfiability problem for* LP$_{CS}$ *is* Σ_2^p-*hard, and the validity problem is* Π_2^p-*hard.*

Together with Corollary 5.26 we conclude the following.

Corollary 5.29. *Let* CS *be a decidable, schematically injective, and axiomatically appropriate constant specification. The satisfiability problem for* LP$_{CS}$ *is* Σ_2^p-*complete, and the validity problem is* Π_2^p-*complete.*

5.3 Notes

NP-completeness of LP$_{CS}$ with a finite constant specification CS can be extracted from Artemov's early papers on justification logic, see, e.g., [5, 6].

The upper complexity bound of Π_2^p for LP$_{CS}$ with a schematic constant specification was demonstrated in [90]. In that paper, Kuznets was working with the *maximal* (sometimes also called *total*) constant specification where each constant justifies every axiom. His proof, however, only relies on the fact that the constant specification is schematic.

Many more complexity results for justification logic are discussed in detail in Kuznets' PhD thesis [92].

The lower bound result given in Theorem 5.28 goes back to Milnikel [110]. Later, Kuznets and Buss [40] reproved it and generalized their method to capture further justification logics.

The notion of a *schematically injective* CS has been introduced by Milnikel [110].

Chapter 6

Arithmetical Interpretation

It is an old question how to provide a classical provability semantics for the modal logic S4 and thus also for intuitionistic logic. In this chapter we first give more details about the mathematical logic tradition for LP$_{CS}$. Then we introduce Peano arithmetic and recall some of its basic proof-theoretic properties such as the Diagonalization Lemma. Making essential use of this lemma, we present two variants of arithmetical interpretations for LP$_{CS}$. Together with the embedding of IPC into S4 and the realization of S4 into LP$_{CS}$, this yields the desired classical provability semantics for intuitionistic logic.

6.1 An Old Problem

The original motivation for developing the first justification logic, the Logic of Proofs, was to provide intuitionistic logic with an adequate provability semantics, i.e., a semantics respecting Brouwer's fundamental idea, see, e.g., [137], that

$$\text{intuitionistic truth means provability.} \tag{6.1}$$

Heyting and Kolmogorov [76, 77, 84] gave an explicit (but informal) definition of this notion of intuitionistic truth, which nowadays is known as Brouwer–Heyting–Kolmogorov (BHK) semantics for intuitionistic logic. It suggests that a formula is intuitionistically true if it has a proof. Further, a proof of a complex formula is given in terms of proofs of its components: namely,

1. a proof of $A \land B$ consists of a proof of A and a proof of B;

2. a proof of $A \vee B$ consists either of a proof of A or of a proof of B;

3. a proof of $A \to B$ is a construction that transforms proofs of A into proofs of B;

4. $\bot$ is a proposition that has no proof.

BHK semantics is widely accepted as the intended semantics for intuitionistic logic. However, it is purely informal and does not provide a precise definition of intuitionistic truth. This was tackled by Gödel [72] who introduced a modal calculus of classical provability (equivalent to S4) with the intended reading of $\Box F$ as *F is provable*. Gödel defined a translation $t(\cdot)$ from IPC to S4 where the translation $t(F)$ of an intuitionistic formula F is given by

prefix each subformula of F with a $\Box$-modality.

He apparently considered this to be an appropriate formalization of Brouwer's thesis (6.1). Gödel established that

$$\mathsf{IPC} \vdash F \quad \text{implies} \quad \mathsf{S4} \vdash t(F)$$

He conjectured that the converse direction also holds, which was later shown by McKinsey and Tarski [108].

However, the ultimate goal of providing a classical provability semantics for IPC was not achieved by Gödel's translation because no precise provability semantics was given for the provability operator $\Box$. The situation was as follows:

$$\mathsf{IPC} \hookrightarrow \mathsf{S4} \hookrightarrow \ldots??? \ldots \hookrightarrow \text{classical proofs} \ .$$

Here, *classical proofs* refers to a system that is based on a formal proof predicate $\mathsf{Proof}(x, y)$ meaning *x is the code of a proof of a formula with the code y* for a classical first-order theory like Peano Arithmetic PA. The problem with this approach, as noted by Gödel himself [72], is that the interpretation of $\Box F$ as the formal provability predicate $\mathsf{Provable}(F) := \exists x \mathsf{Proof}(x, F)$ is incompatible with arithmetical reasoning. Indeed, the formula $\Box \bot \to \bot$, an axiom of S4, is interpreted as

$$\neg \mathsf{Provable}(\bot) \ .$$

In other words,

it is not the case that $\bot$ is provable in PA,

6.1 An Old Problem

or, equivalently

$$\text{PA is consistent.}$$

Hence, by modal necessitation we obtain the S4-theorem $\Box(\Box\bot \to \bot)$, interpreted as

"PA is consistent" is provable in PA,

in direct contradiction to Gödel's second incompleteness theorem.

So there was the provability calculus S4 without a formal classical provability semantics and a classical provability interpretation of $\Box$ as Provable($\cdot$) without a corresponding axiom system. Thus, Gödel left open two natural problems:

1. find the modal logic of the formal provability predicate Provable($\cdot$);

2. find a formal provability semantics for S4 and thereby for IPC.

The first problem lead to the highly successful field of provability logic [11, 30, 80, 126]. The problem itself was solved by showing that a system called GL (for Gödel–Löb) provides a complete axiomatization of the logic of formal provability [127].

Artemov [4, 6] solved the second problem by introducing the Logic of Proofs. As shown above, the problem with finding a provability semantics for S4 is to interpret the reflection principle $\Box F \to F$, since

$$\exists x \mathsf{Proof}(x, \bot) \to \bot$$

is not provable in PA. The point is that $\exists x \mathsf{Proof}(x, F)$ does not yield a specific proof of F since x may stand for a non-standard natural number, which does not represent the code of any actual proof in PA. The situation is completely different for proofs represented by explicit terms. Namely, the principle

$$\mathsf{Proof}(p, F) \to F \tag{6.2}$$

is provable for each specific code p of a proof. Indeed, if $\mathsf{Proof}(p, F)$ is true, then F is provable in PA, and so is (6.2). Otherwise, if $\mathsf{Proof}(p, F)$ is false, then $\neg \mathsf{Proof}(p, F)$ is true and hence provable[1], which again implies that (6.2) is provable. This observation suggests that a provability

---

[1] Technically speaking, this conclusion follows the completeness of PA with respect to the standard model for "sufficiently simple" formulas and the fact that Proof can be represented by such a simple formula. We provide the necessary technical details later in this chapter.

semantics for S4 can be obtained by first interpreting S4 into a system with explicit proof terms and then providing a provability semantics for that system.

The Logic of Proofs $\mathsf{LP_{CS}}$, developed by Artemov, is such a system with explicit proof terms. Artemov showed that it gives a provability semantics for S4. From the Realization Theorem (Theorem 3.22) we know that there is an interpretation of S4 into $\mathsf{LP_{CS}}$ (for axiomatically appropriate and schematic CS). Further, we show in this chapter that there is a formal provability semantics for $\mathsf{LP_{CS}}$, which gives us the following chain of exact embeddings:

$$\mathsf{IPC} \hookrightarrow \mathsf{S4} \hookrightarrow \mathsf{LP_{CS}} \hookrightarrow \text{classical proofs} .$$

Hence the Logic of Proofs can be viewed as a formalization of the BHK semantics for intuitionistic propositional logic.

In the following, we first recall some basic facts about Peano Arithmetic. Then we introduce a *simple arithmetical interpretation* for $\mathsf{LP_{CS}}$ and show that for almost schematic constant specifications CS, the logic $\mathsf{LP_{CS}}$ is sound and complete with respect to simple arithmetical interpretations. However, simple arithmetical interpretations do not provide a semantics for the term operations $\cdot$, $+$, and ! as computable functions working on all codes of proofs. This is remedied in Artemov's original *arithmetical interpretation*, which provides a computational semantics for the term operations. The price to pay for this additional feature is that we have arithmetical interpretations only for finite constant specifications. We establish that $\mathsf{LP_{CS}}$ is sound and complete with respect to arithmetical interpretations if CS if finite.

6.2 Preliminaries

In this section we introduce all notions and concepts of primitive recursive functions, recursive functions, and Peano Arithmetic that we later need to present arithmetical interpretations for the Logic of Proofs. Readers who are familiar with Peano Arithmetic, in particular with formal proof predicates and diagonalization arguments, may safely skip this section.

6.2.1 Primitive Recursive Functions

We start with recalling the primitive recursive functions and relations on the natural numbers $\mathbb{N}$. Throughout this section we use $\vec{k}$ for a

6.2 Preliminaries

sequence of natural numbers $k_0, \ldots, k_{n-1}$. To avoid treating trivial cases separately, we also use the notation $f(\vec{k})$ when $n = 0$, i.e., when $\vec{k}$ is the empty list and f is a nullary function. In such cases $f(\vec{k})$ is to be understood as f.

Definition 6.1 (Primitive recursive function). The class of *primitive recursive functions* is inductively defined as follows.

1. The following functions on $\mathbb{N}$ are primitive recursive:

   (a) the unary successor function $\mathsf{s}(k) := k + 1$,

   (b) for each $n \in \mathbb{N}$, the n-ary constant functions $\mathsf{cst}_m^n(\vec{k}) := m$ for each natural number $m \in \mathbb{N}$,[2] and

   (c) for each $n \in \mathbb{N}$ such that $n \geq 1$, the n-ary projection functions $\mathsf{pr}_i^n(\vec{k}) := k_i$ for each $i \in \mathbb{N}$ such that $0 \leq i < n$.

2. If $f : \mathbb{N}^m \to \mathbb{N}$ is an m-ary and $g_1, \ldots, g_m : \mathbb{N}^n \to \mathbb{N}$ are n-ary primitive recursive functions for some $m, n \in \mathbb{N}$, then the n-ary composition

$$\bigl[\mathsf{Comp}(f, g_1, \ldots, g_m)\bigr](\vec{k}) := f\bigl(g_1(\vec{k}), \ldots, g_m(\vec{k})\bigr)$$

   of f and $g_1, \ldots, g_m$ is primitive recursive.

3. If $f : \mathbb{N}^n \to \mathbb{N}$ and $g : \mathbb{N}^{n+2} \to \mathbb{N}$ are an n-ary and an $(n+2)$-ary primitive recursive functions respectively, then the $(n+1)$-ary function $\mathsf{Rec}(f, g) : \mathbb{N}^{n+1} \to \mathbb{N}$ that is obtained from f and g by applying the following schema of primitive recursion:

$$\bigl[\mathsf{Rec}(f, g)\bigr](0, \vec{k}) := f(\vec{k}) ,$$
$$\bigl[\mathsf{Rec}(f, g)\bigr](l+1, \vec{k}) := g\Bigl(\bigl[\mathsf{Rec}(f, g)\bigr](l, \vec{k}), l, \vec{k}\Bigr) ,$$

   is primitive recursive.

Hence, the class of primitive recursive functions is the smallest class of functions that contains s, cst_m^n for all $n, m \in \mathbb{N}$, and pr_i^n for all $n, i \in \mathbb{N}$ such that $n \geq 1$ and $0 \leq i < n$ and that is closed under composition and the schema of primitive recursion.

---
[2] This includes constants for $n = 0$.

Now we present several primitive recursive functions, which will be used later. The proofs that these functions are indeed primitive recursive are completely standard and can be found in many textbooks, see, e.g., [69, 109].

Binary addition, multiplication, exponentiation (with $0^0 := 1$), and the *natural subtraction* function, given by

$$k \mathbin{\dot{-}} l := \begin{cases} 0 & \text{if } k < l, \\ k - l & \text{otherwise}, \end{cases}$$

are primitive recursive, as well as the unary **sign** function:

$$\mathsf{sign}(k) := \begin{cases} 0 & \text{if } k = 0, \\ 1 & \text{otherwise}. \end{cases}$$

The factorial function with $0! = 1$ and $(k+1)! = (k!) \cdot (k+1)$ also is primitive recursive. If $f(l, \vec{k})$ is an $(n+1)$-ary primitive recursive function, so is the *bounded product* function $[\prod f]$ based on f

$$\left[\prod f\right](m, \vec{k}) := \prod_{l < m} f(l, \vec{k}) \ .$$

Note that $[\prod f](0, \vec{k}) = 1$ as the empty product for any sequence $\vec{k}$ of natural numbers. The unary function $p(i)$ with

$$p(i) := i\text{th prime number}$$

is primitive recursive (where we start with $p(0) := 2$). Indeed, we can define $p(i+1)$ to be the least natural number $k \leq p(i)! + 1$ such that

$$p(i) < k \land k \text{ is prime},$$

which can be expressed using primitive recursive functions. To simplify the notation, as is commonly done, we will often write p_i instead of $p(i)$.

An important property of primitive recursive functions is that they can be used to encode finite sequences of natural numbers. For instance, for each $n \in \mathbb{N}$, the n-ary function $\langle \cdot, \ldots, \cdot \rangle : \mathbb{N}^n \to \mathbb{N}$ defined by

$$\langle k_0, \ldots, k_{n-1} \rangle := \prod_{i < n} p_i^{k_i + 1}$$

6.2 Preliminaries

is primitive recursive. For instance, it encodes the sequence $4, 6, 0$ by the natural number

$$\langle 4, 6, 0 \rangle = 2^{4+1} \cdot 3^{6+1} \cdot 5^{0+1} \ .$$

Note that the constant $\langle \rangle$, the encoding for the empty sequence, is another name for the natural number 1. Further, there is[3] a unary primitive recursive function $\mathsf{len} : \mathbb{N} \to \mathbb{N}$ that satisfies

$$\mathsf{len}(\langle \rangle) = 0 \quad \text{and} \quad \mathsf{len}(\langle k_0, \ldots, k_{n-1} \rangle) = n \ .$$

To call this encoding primitive recursive, not only the coding functions $\langle \cdot, \ldots, \cdot \rangle$ but also the corresponding decoding function should be primitive recursive: an element of a coded finite sequence of natural numbers can be accessed by a binary primitive recursive function $\mathsf{proj} : \mathbb{N}^2 \to \mathbb{N}$ such that[4]

$$\mathsf{proj}(\langle k_0, \ldots, k_{n-1} \rangle, i) = k_i$$

for $0 \leq i < n$. Instead of $\mathsf{proj}(l, i)$ we often write $(l)_i$. Moreover, concatenation of two finite sequences of natural numbers is primitive recursive: there is a binary primitive recursive function $\mathsf{concat} : \mathbb{N}^2 \to \mathbb{N}$ such that

$$\mathsf{concat}(\langle k_0, \ldots, k_{m-1} \rangle, \langle l_0, \ldots, l_{n-1} \rangle) = \langle k_0, \ldots, k_{m-1}, l_0, \ldots, l_{n-1} \rangle \ .$$

Definition 6.2 (Relation). An n-ary relation R on $\mathbb{N}$ is a subset $R \subseteq \mathbb{N}^n$.[5]

We assume that each relation is supplied with its arity. However, in most cases, the arity of R can be easily inferred from the elements of R. (For instance, $\{\varnothing\}$ can only be a nullary relation.) The only exception is empty relations. We distinguish the nullary, unary, binary, ternary, etc. empty relations on $\mathbb{N}$ and denote them by $\varnothing^{(0)}$, $\varnothing^{(1)}$, $\varnothing^{(2)}$, $\varnothing^{(3)}$,

---

[3] The existential quantifier here is due to the fact that primitive recursive functions are always total, but not all natural numbers are encodings of finite sequences under $\langle \cdot, \ldots, \cdot \rangle$. For instance, neither 0 nor $2 \cdot 5$ encodes anything. Thus, our definition of len is incomplete. Since we do not care about the values of len on non-codes, we do not specify the remaining values but claim instead that they can be specified in such a way that the resulting function is primitive recursive.

[4] Similar to len, we do not specify the value of $\mathsf{proj}(\langle \rangle, i)$ and, more generally, what happens when $i \geq n$.

[5] This includes Boolean constants for $n = 0$, where $\mathbb{N}^0 := \{\varnothing\}$. In particular, we can identify $\varnothing$ with *false* and $\{\varnothing\}$ with *true*.

etc. Instead of $(k_0, \ldots, k_{n-1}) \in R$ we usually write $R(k_0, \ldots, k_{n-1})$, once again using the same notation even when $n = 0$ to mean simply R. Since $\varnothing$ is the only element that may or may not belong to a nullary relation, by R we then understand the statement $\varnothing \in R$, which is true for $R = \{\varnothing\}$ and false for $R = \varnothing^{(0)}$.

Definition 6.3 (Characteristic function)**.** We define the *characteristic function* χ_R of an n-ary relation R by

$$\chi_R(k_0, \ldots, k_{n-1}) := \begin{cases} 0 & \text{if } R(k_0, \ldots, k_{n-1}) \ , \\ 1 & \text{otherwise} \ . \end{cases}$$

Thus, the characteristic function of an n-ary relation on $\mathbb{N}$ is an n-ary function on $\mathbb{N}$.

Note that the characteristic functions of the empty relations of different arity are distinct: $\chi_{\varnothing^{(n)}} = \mathsf{cst}_1^n$. They are all constant 1 functions but with different domains. This is one of the reasons different empty relations need to be distinguished.

Definition 6.4 (Primitive recursive relation)**.** An n-ary relation is called *primitive recursive* if its characteristic function is primitive recursive.

Note that the usual less-than relation $<$ as well as equality $=$ are binary primitive recursive relations.

Primitive recursive relations are closed under unions, intersections, complementation, and under bounded quantification:

Lemma 6.5. *Let $\vec{k}$ be $k_0, \ldots, k_{n-1}$.*

1. *The n-ary empty relation is primitive recursive.*

2. *Let R and S be n-ary primitive recursive relations. Then the n-ary relations $R \cup S$, $R \cap S$,*

$$\overline{R} := \mathbb{N}^n \setminus R, \quad \text{and} \quad R \to S := \overline{R} \cup S$$

   are primitive recursive.

3. *Let R be an $(n+1)$-ary primitive recursive relation. Then the $(n+1)$-ary relations*

$$R_\exists := \{(l, \vec{k}) \mid (\exists m < l) R(m, \vec{k})\}$$

   and

$$R_\forall := \{(l, \vec{k}) \mid (\forall m < l) R(m, \vec{k})\}$$

   are primitive recursive.

6.2 Preliminaries

4. Let $R_0, \ldots, R_{l-1}$ be n-ary pairwise disjoint primitive recursive relations and $f_0, \ldots, f_l$ be n-ary primitive recursive functions for some $l \geq 1$. Then the following function is primitive recursive:

$$f(\vec{k}) := \begin{cases} f_0(\vec{k}) & \text{if } R_0(\vec{k}), \\ \cdots & \cdots \\ f_{l-1}(\vec{k}) & \text{if } R_{l-1}(\vec{k}), \\ f_l(\vec{k}) & \text{otherwise.} \end{cases}$$

Proof.

1. As already mentioned earlier, $\chi_{\emptyset^{(n)}} = \mathsf{cst}_1^n$, the n-ary constant 1 function, which is primitive recursive by definition.

2. Because all constants are nullary primitive recursive functions, both nullary relations on $\mathbb{N}$ are primitive recursive. It remains to consider relations of positive arity.

   The characteristic functions of $R \cup S$ and $\overline{R}$

   $$\chi_{R \cup S} = \chi_R \cdot \chi_S ,$$
   $$\chi_{\overline{R}} = \mathsf{cst}_1^n \dotminus \chi_R$$

   are primitive recursive as compositions of primitive recursive functions of n-ary constant 1, natural subtraction, multiplication, and the characteristic functions χ_R and χ_S. Further, $R \cap S$ and $R \to S$ are primitive recursive because they can be represented through union and complementation, $R \to S$ by its definition and

   $$R \cap S = \overline{\overline{R} \cup \overline{S}} .$$

3. The characteristic function of $R_\exists$

   $$\chi_{R_\exists}(l, \vec{k}) = \prod_{m < l} \chi_R(m, \vec{k}) ,$$

   is primitive recursive as a bounded product based on the characteristic function χ_R, which is primitive recursive by assumption.

   It remains to show that $R_\forall$ is primitive recursive. This follows now easily from the following representation

   $$R_\forall = \overline{\overline{R}_\exists}$$

   together with the closure conditions previously established in this lemma.

4. It is easy to see that
$$f = f_0 \cdot \chi_{\overline{R_0}} + \cdots + f_{l-1} \cdot \chi_{\overline{R_{l-1}}} + f_l \cdot \chi_{R_0 \cup \cdots \cup R_{l-1}} .$$

Since $\overline{R_0}, \ldots, \overline{R_{l-1}}$, and $R_0 \cup \cdots \cup R_{l-1}$ are primitive recursive relations as previously established in this lemma, it follows that their characteristic functions are primitive recursive. □

Corollary 6.6 (Finite relations are primitive recursive). *If R is an n-ary relation on $\mathbb{N}$ such that R is finite as a set, then R is primitive recursive.*

Proof. Given that the empty relation is primitive recursive and that any finite union of primitive recursive relations is primitive recursive, it is sufficient to show that any singleton set R is primitive recursive. Let $R = \{(m_0, \ldots, m_{n-1})\}$. Then it is easy to see that, for $\vec{k} = k_0, \ldots, k_{n-1}$,

$$\chi_R(\vec{k}) = \mathsf{sign}\left(\left(\mathsf{pr}_0^n(\vec{k}) \mathbin{\dot{-}} \mathsf{cst}_{m_0}^n(\vec{k}) \right)^2 + \cdots + \left(\mathsf{pr}_{n-1}^n(\vec{k}) \mathbin{\dot{-}} \mathsf{cst}_{m_{n-1}}^n(\vec{k}) \right)^2 \right) .$$

□

Remark 6.7. The definition of χ_R above may be a bit hard to parse. A working mathematician would write it as

$$\chi_R(\vec{k}) = \mathsf{sign}\left((k_0 \mathbin{\dot{-}} m_0)^2 + \cdots + (k_{n-1} \mathbin{\dot{-}} m_{n-1})^2 \right) .$$

What causes the complexity is that the definitions of composition and primitive recursion demand all the input arguments to be recited every time, even when constants are used. Since the not-fully-correct version is much more transparent for human eye, in the rest of this chapter we avoid the verbose correct versions, using the fact that they can easily be restored.

6.2.2 Recursive Functions

As mentioned earlier, primitive recursive functions are total as follows easily from their definition, i.e., the domain of any n-ary primitive recursive function is $\mathbb{N}^n$. We now relax this restriction by defining the classes of n-ary *recursive* functions, which can be partial, i.e., whose domain is any subset of $\mathbb{N}^n$, i.e., $f(\vec{k})$ may be undefined for a partial recursive function f and any particular values $\vec{k}$. We write $f(\vec{k}){\uparrow}$ to express that

$f(\vec{k})$ is undefined, with the intended computational interpretation that the corresponding computation does not terminate. We write $f(\vec{k})\downarrow$ to express that $f(\vec{k})$ is defined. One has to be precise when talking about equality of such partial functions. Apart from requiring the functions to be equal when both are defined, one must also prohibit the situation when one function is defined and the other is not.

Definition 6.8 (Equality of partial functions). Two partial n-ary natural number functions f and g are called equal, written $f \simeq g$, if for any sequence of numbers $\vec{k}$ of length n either

$f(\vec{k})\uparrow$ and $g(\vec{k})\uparrow$ or

$f(\vec{k})\downarrow$, $g(\vec{k})\downarrow$, and $f(\vec{k}) = g(\vec{k})$.

Definition 6.9 (Recursive function). The class of *recursive functions*, or *computable functions*, is inductively defined by:

1. The functions s, cst_m^n for all $n, m \in \mathbb{N}$, and pr_i^n for all $n, i \in \mathbb{N}$ such that $n \geq 1$ and $0 \leq i < n$ are recursive. Their arity is, of course, the same as in the definition of primitive recursive functions.

2. The recursive functions are closed under composition Comp.

3. The recursive functions are closed under primitive recursion Rec.

4. The recursive functions are closed under the μ-schema: if

$$f : \mathbb{N}^{n+1} \to \mathbb{N}$$

is an $(n+1)$-ary recursive function, then also the n-ary function

$$\mu.f : \mathbb{N}^n \to \mathbb{N}$$

given by

$$[\mu.f](\vec{k}) := \begin{cases} l & \text{if } f(l, \vec{k}) = 0 \wedge \\ & \quad (\forall m < l)(f(m, \vec{k})\downarrow \wedge f(m, \vec{k}) > 0) \ , \\ \text{undefined} & \text{otherwise} \end{cases}$$

is recursive. Hence, informally $[\mu.f](\vec{k})$ represents the following algorithm for computing the least natural number l such that $f(l, \vec{k}) = 0$: first $f(0, \vec{k})$ is computed, then $f(1, \vec{k})$, then $f(2, \vec{k})$, etc., until the function f returns 0 for the first time. If this algorithm does not terminate either because the computation of $f(m, \vec{k})$ does not terminate for some m or because f never returns 0, then $\mu.f$ is considered undefined.

A recursive function that is total is called a *total recursive function*.

Example 6.10. $\mu.\mathsf{cst}_1^1$ is a nullary recursive function that is undefined.

Remark 6.11. Since both function composition Comp and primitive recursion Rec applied to total functions yield a total function, the class of total recursive functions is closed under Comp and Rec. At the same time, as Example 6.10 shows, this class is not closed under the μ-schema.

Remark 6.12. Since the function f defined by cases in Lemma 6.5(4) is obtained by using primitive recursive operations applied to f_i's and the characteristic functions of various relations, it follows that for recursive functions f_i's and primitive recursive relations R_i's, such a function f is recursive. Moreover, it is easy to see that f is total recursive if all f_i's are.

6.2.3 Peano Arithmetic

Definition 6.13. The *language $\mathcal{L}_{\mathsf{PA}}$ of arithmetic* is the language of first-order logic with countably many *individual variables*, the logical symbols $\bot, \rightarrow, \forall$, and the following non-logical symbols:

1. an n-ary function symbol $\underline{f}$ for each n-ary primitive recursive function f (this includes binary function symbols $\underline{+}$ and $\underline{\cdot}$ and a unary function symbol $\underline{s}$);

2. an n-ary relation symbol $\underline{R}$ for each n-ary primitive recursive relation R, including a binary relation symbol $\underline{=}$.

We use $x, y, z, \ldots$ with or without sub and/or superscripts to denote individual variables of $\mathcal{L}_{\mathsf{PA}}$ and hope the reader is able to distinguish them from the justification variables of $\mathcal{L}_{\mathsf{J}}$. Further, we denote formulas of $\mathcal{L}_{\mathsf{PA}}$ by $\phi, \psi, \ldots$ with or without sub and/or superscripts. Formally, *terms of $\mathcal{L}_{\mathsf{PA}}$* are defined as follows

1. Every variable x is a term of $\mathcal{L}_{\mathsf{PA}}$;

6.2 Preliminaries

2. The nullary function symbol $\underline{\mathsf{cst}^0_m}$ is a term of $\mathcal{L}_{\mathsf{PA}}$ for each $m \in \mathbb{N}$;[6]

3. If $t_0, \ldots, t_{n-1}$ are terms of $\mathcal{L}_{\mathsf{PA}}$ for $n \geq 1$, then so is $\underline{f}(t_0, \ldots, t_{n-1})$ for each n-ary primitive recursive function f.[7]

Formulas of $\mathcal{L}_{\mathsf{PA}}$ are defined as follows:

1. $\bot$ is a formula of $\mathcal{L}_{\mathsf{PA}}$;

2. A nullary relation symbol $\underline{R}$ is a formula of $\mathcal{L}_{\mathsf{PA}}$ for each nullary primitive recursive relation R;

3. If $t_0, \ldots, t_{n-1}$ are terms of $\mathcal{L}_{\mathsf{PA}}$ for $n \geq 1$, then $\underline{R}(t_0, \ldots, t_{n-1})$ is a formula of $\mathcal{L}_{\mathsf{PA}}$ for each n-ary primitive recursive relation R;[8]

4. If ϕ and ψ are formulas of $\mathcal{L}_{\mathsf{PA}}$, then so is $(\phi \to \psi)$;

5. If ϕ is a formula of $\mathcal{L}_{\mathsf{PA}}$, then so is $\forall x \phi$ for each variable x.

We do not provide more details here because they are rather standard and would take a lot of space. In particular, *free* and *bound* occurrences of variables in a formula are defined in the usual way. A *sentence* is a formula without free occurrences of variables; a *ground term* is a term without (free) occurrences of variables. As usual $\exists x \phi := \neg \forall x \neg \phi$. Further, we define bounded quantifiers $(\forall x < t)\phi := \forall x(x < t \to \phi)$ and $(\exists x < t)\phi := \exists x(x < t \wedge \phi)$, where t may not contain (free) occurrences of x.

Definition 6.14 (Assignment, value of a term). A *natural assignment*, or simply an *assignment*[9], is a mapping v that assigns $\mathsf{v}(x) \in \mathbb{N}$ to each variable x of $\mathcal{L}_{\mathsf{PA}}$. Given an assignment v, the value $t^{\mathsf{v}} \in \mathbb{N}$ of an $\mathcal{L}_{\mathsf{PA}}$-term t is inductively given by

$$x^{\mathsf{v}} := \mathsf{v}(x) \quad \text{and} \quad \left(\underline{f}(t_0, \ldots, t_{n-1})\right)^{\mathsf{v}} := f(t_0^{\mathsf{v}}, \ldots, t_{n-1}^{\mathsf{v}})$$

---

[6] Note that each nullary primitive recursive function has the form cst^0_m for some natural number m because these functions are defined and constant. Needless to say, many constructions using Comp and Rec produce the same nullary function, but they all correspond to the same nullary function symbol that we denote $\underline{\mathsf{cst}^0_m}$.

[7] Strictly speaking, the case of $\underline{\mathsf{cst}^0_m}$ could be subsumed by this case. We separated the former to emphasize the two types of atomic terms: variables and constant symbols.

[8] Once again, the case of nullary relation symbols could be subsumed by this case.

[9] We are primarily interested in the standard model $\mathbb{N}$ of arithmetic. More generally, assignments assign to variables values from an arbitrary chosen domain.

for each variable x and each n-ary primitive recursive function f. In particular, $\left(\underline{\mathsf{cst}_m^0}\right)^{\mathsf{v}} := m$ for each $m \in \mathbb{N}$.

An immediate consequence of this definition is that for any ground $\mathcal{L}_{\mathsf{PA}}$-term t, we have $t^{\mathsf{v}_1} = t^{\mathsf{v}_2}$ for arbitrary assignments v_1 and v_2. For a ground term t, the natural number that is the common value of t^{v} for all assignments v is called the $\mathbb{N}$-*value of* t and denoted by $t^{\mathbb{N}}$. In particular, each natural number m is the $\mathbb{N}$-value of the ground term $\underline{\mathsf{cst}_m^0}$, i.e., $\left(\underline{\mathsf{cst}_m^0}\right)^{\mathbb{N}} = m$. Hence, we can consider the ground term $\underline{\mathsf{cst}_m^0}$ to be the canonical representation of the natural number m in $\mathcal{L}_{\mathsf{PA}}$. We call terms $\underline{\mathsf{cst}_m^0}$ *numerals*. To simplify the notation in $\mathcal{L}_{\mathsf{PA}}$, we will often write $\underline{m}$ instead of $\underline{\mathsf{cst}_m^0}$.

Similarly, when working in $\mathcal{L}_{\mathsf{PA}}$ we will often use 0, s, $+$, $\cdot$, $=$, and $<$ for $\underline{0}$, $\underline{\mathsf{s}}$, $\underline{+}$, $\underline{\cdot}$, $\underline{=}$ and $\underline{<}$ because the latter are often used as primary symbols for Peano Arithmetic. We also use infix notation for $+$, $\cdot$, $=$, and $<$ and omit underlining for functions like $\langle \cdot, \ldots, \cdot \rangle$, $(\cdot)_i$, etc.

Another notational convention concerns substitutions. Writing

$$[x \mapsto t]$$

as before is a bit too clumsy for first-order logic because substitution here is a logical rather than metalogical operation. Accordingly, we will write $\phi(x)$ to point to the variable whose free occurrences are to be replaced with various terms. The notation $\phi(x)$ implies neither that x has free occurrences in the formula ϕ nor that it is the only variable occurring free in ϕ. It simply means that $\phi(t)$ should be understood as $\phi[x \mapsto t]$. The same goes for $\phi(x_0, \ldots, x_{n-1})$ and $\phi(t_0, \ldots, t_{n-1})$. As usual in first-order languages, we always implicitly assume that each term t_i is free for the variable x_i within ϕ, i.e., no free occurrence of x_i in ϕ is under the scope of a quantifier over any variable occurring in t_i. If this condition is not satisfied, the bound variables in ϕ need to be renamed. We will not discuss this explicitly because it is part of core knowledge about first-order logic.

Definition 6.15 (Truth in standard model). We define when a formula ϕ of $\mathcal{L}_{\mathsf{PA}}$ is *true under an assignment* v (written $(\mathbb{N}, \mathsf{v}) \vDash \phi$) inductively by:

1. $(\mathbb{N}, \mathsf{v}) \nvDash \bot$;

6.2 Preliminaries

2. $(\mathbb{N}, \mathsf{v}) \vDash \underline{R}(t_0, \ldots, t_{n-1})$ iff $\chi_R(t_0^{\mathsf{v}}, \ldots, t_{n-1}^{\mathsf{v}}) = 0$ for any n-ary primitive recursive relation R and for arbitrary terms $t_0, \ldots, t_{n-1}$;

3. $(\mathbb{N}, \mathsf{v}) \vDash \phi \to \psi$ iff $(\mathbb{N}, \mathsf{v}) \nvDash \phi$ or $(\mathbb{N}, \mathsf{v}) \vDash \psi$;

4. $(\mathbb{N}, \mathsf{v}) \vDash \forall x \phi$ iff $(\mathbb{N}, \mathsf{v}) \vDash \phi[x \mapsto \underline{m}]$ for all $m \in \mathbb{N}$.[10]

If ϕ is a sentence, we obviously have $(\mathbb{N}, \mathsf{v}_1) \vDash \phi$ if and only if $(\mathbb{N}, \mathsf{v}_2) \vDash \phi$ for arbitrary assignments v_1 and v_2. In this case we write $\mathbb{N} \vDash \phi$ and say that ϕ is true.

It immediately follows from this definition that

1. $(\mathbb{N}, \mathsf{v}) \vDash \exists x \phi$ iff $(\mathbb{N}, \mathsf{v}) \vDash \phi[x \mapsto \underline{m}]$ for some $m \in \mathbb{N}$;

2. $(\mathbb{N}, \mathsf{v}) \vDash (\forall x < t)\phi$ iff $(\mathbb{N}, \mathsf{v}) \vDash \phi[x \mapsto \underline{m}]$ for all $m \in \mathbb{N}$ such that $m < t^{\mathsf{v}}$;

3. $(\mathbb{N}, \mathsf{v}) \vDash (\exists x < t)\phi$ iff $(\mathbb{N}, \mathsf{v}) \vDash \phi[x \mapsto \underline{m}]$ for some $m \in \mathbb{N}$ such that $m < t^{\mathsf{v}}$.

Peano Arithmetic PA is given in the language $\mathcal{L}_{\mathsf{PA}}$. It comprises the axioms and rules of first-order predicate logic, equality axioms for the binary relation symbol $=$, and axioms for all primitive recursive functions and relations. We show only the non-logical axioms of PA, which formalize the primitive recursive functions and relations.

Equality axioms:

$x = x$;

$x = y \to (\phi \to \phi[x \mapsto y])$ for any formula $\phi \in \mathcal{L}_{\mathsf{PA}}$ and arbitrary variables x and y.

Axioms for the successor function:

$\neg(0 = \mathsf{s}(x))$;

$\mathsf{s}(x) = \mathsf{s}(y) \to x = y$;

$\mathsf{s}(\underline{n}) = \underline{n+1}$ for each $n \in \mathbb{N}$.

---

[10]The case for $\forall$ only works in the standard model. For more general models of arithmetic, truth for $\forall x \phi$ must be defined by considering different values of $\mathsf{v}(x)$ within the domain because not all non-standard numbers can be represented by terms.

Axioms for the constant functions: for arbitrary $n, m \in \mathbb{N}$,
$$\underline{\mathsf{cst}^n_m}(x_0, \ldots, x_{n-1}) = \underline{m}.$$

Axioms for the projection functions: for arbitrary $n, i \in \mathbb{N}$, such that $n \geq 1$ and $0 \leq i < n$,
$$\underline{\mathsf{pr}^n_i}(x_0, \ldots, x_{n-1}) = x_i.$$

Axioms for function composition: for arbitrary m-ary primitive recursive function f and n-ary primitive recursive functions $g_1, \ldots, g_m$,
$$\underline{\mathsf{Comp}(f, g_1, \ldots, g_m)}(x_0, \ldots, x_{n-1}) =$$
$$\underline{f}\Big(\underline{g_1}(x_0, \ldots, x_{n-1}), \ldots, \underline{g_m}(x_0, \ldots, x_{n-1})\Big).$$

Axioms for primitive recursion: for arbitrary n-ary primitive recursive function f and $(n+2)$-ary primitive recursive function g,
$$\underline{\mathsf{Rec}(f, g)}(0, x_0, \ldots, x_{n-1}) = \underline{f}(x_0, \ldots, x_{n-1});$$
$$\underline{\mathsf{Rec}(f, g)}\big(\mathsf{s}(y), x_0, \ldots, x_{n-1}\big) =$$
$$\underline{g}\Big(\underline{\mathsf{Rec}(f, g)}(y, x_0, \ldots, x_{n-1}), \, y, \, x_0, \, \ldots, \, x_{n-1}\Big).$$

Axioms for primitive recursive relations: for any n-ary primitive recursive relation
$$\underline{R}(x_0, \ldots, x_{n-1}) \leftrightarrow \underline{\chi_R}(x_0, \ldots, x_{n-1}) = 0.$$

Induction axioms: for any formula $\phi \in \mathcal{L}_{\mathsf{PA}}$,
$$\phi(0) \wedge \forall x \Big(\phi(x) \to \phi\big(\mathsf{s}(x)\big)\Big) \to \forall x \phi(x).$$

As usual, we write $\mathsf{PA} \vdash \phi$ if the formula ϕ is provable in PA.

Remark 6.16. Our formulation of Peano Arithmetic includes terms for all primitive recursive functions and relations. Often Peano Arithmetic is given as a system PA' in a smaller language $\mathcal{L}'_{\mathsf{PA}}$ that only includes function symbols $\underline{\mathsf{s}}$ for the successor function, $\underline{+}$ for addition, and $\underline{\cdot}$ for multiplication, a constant symbol $\underline{0}$, i.e., $\underline{\mathsf{cst}^0_0}$, for zero, and a relation symbol $\underline{=}$ for equality.

It is a classic result, see, e.g., [69], that PA is a conservative extension of PA'. More precisely, there exists a translation $(\cdot)' \colon \mathcal{L}_{\mathsf{PA}} \to \mathcal{L}'_{\mathsf{PA}}$ such that for each formula $\phi \in \mathcal{L}_{\mathsf{PA}}$

6.2 Preliminaries

1. $\phi' = \phi$ for each $\phi \in \mathcal{L}'_{\mathsf{PA}}$;

2. ϕ' has the same free variables as ϕ;

3. $\mathsf{PA} \vdash \phi \leftrightarrow \phi'$;

4. $\mathsf{PA} \vdash \phi$ if and only if $\mathsf{PA}' \vdash \phi'$.

Obviously, the axioms of PA are chosen in such a way that they are true in the standard model. That means we assume soundness of PA with respect to $\mathbb{N}$: for all $\mathcal{L}_{\mathsf{PA}}$-sentences ϕ

$$\mathsf{PA} \vdash \phi \quad \text{implies} \quad \mathbb{N} \vDash \phi \ . \tag{6.3}$$

Let us show that for each primitive recursive function f, the corresponding function symbol $\underline{f}$ acts on numerals within the formal system PA exactly like the function itself acts on natural numbers and that this identical behavior is provable within PA.

Lemma 6.17. *Let f be an n-ary primitive recursive function and*

$$m_0, \ldots, m_{n-1}$$

be natural numbers. Using abbreviations $\vec{m}$ for $m_0, \ldots, m_{n-1}$ and $\underline{\vec{m}}$ for $\underline{m_0}, \ldots, \underline{m_{n-1}}$, we have

$$\mathsf{PA} \vdash \underline{f}\left(\underline{\vec{m}}\right) = \underline{f(\vec{m})} \ .$$

Proof. We proceed by induction on the structure of the primitive recursive function f. We distinguish the following cases:

1. If $f = \mathsf{cst}^n_l$, then $\mathsf{cst}^n_l(\vec{m}) = l$.[11] By using the axioms for the constant functions and first-order reasoning, we have

$$\mathsf{PA} \vdash \underline{\mathsf{cst}}^n_l\left(\underline{\vec{m}}\right) = \underline{l} \ .$$

2. If $f = \mathsf{s}$, then $\mathsf{s}(m_0) = m_0 + 1$. By the axioms for the successor function, we have $\mathsf{PA} \vdash \underline{\mathsf{s}}\left(\underline{m_0}\right) = \underline{m_0 + 1}$.

---

[11] As mentioned before, this case contains all nullary primitive recursive function. However, for them, the statement is trivial and can also be derived from the equality axioms.

3. If $f = \mathsf{pr}_l^n$ with $0 \leq l < n$, then $\mathsf{pr}_l^n(\vec{m}) = m_l$. By the axioms for the projection functions and first-order reasoning, $\mathsf{PA} \vdash \underline{\mathsf{pr}_l^n}\left(\underline{\vec{m}}\right) = \underline{m_l}$.

4. Let $f = \mathsf{Comp}(h, g_1, \ldots, g_l)$ for some l-ary primitive recursive function h and some n-ary primitive recursive functions $g_1, \ldots, g_l$. By the induction hypothesis for $h, g_1, \ldots, g_l$,

$$\mathsf{PA} \vdash \underline{h}\left(\underline{g_1(\vec{m})}, \ldots, \underline{g_l(\vec{m})}\right) = \underline{h(g_1(\vec{m}), \ldots, g_l(\vec{m}))}, \quad (6.4)$$

$$\mathsf{PA} \vdash \underline{g_i}\left(\underline{\vec{m}}\right) = \underline{g_i(\vec{m})} \quad \text{for each } 1 \leq i \leq l. \quad (6.5)$$

Since $[\mathsf{Comp}(h, g_1, \ldots, g_l)](\vec{m}) = h(g_1(\vec{m}), \ldots, g_l(\vec{m}))$, we need to show that

$$\mathsf{PA} \vdash \underline{\mathsf{Comp}(h, g_1, \ldots, g_l)}\left(\underline{\vec{m}}\right) = \underline{h(g_1(\vec{m}), \ldots, g_l(\vec{m}))},$$

which follows by first-order reasoning from (6.4), (6.5), and the axioms for function composition that guarantee

$$\mathsf{PA} \vdash \underline{\mathsf{Comp}(h, g_1, \ldots, g_l)}\left(\underline{\vec{m}}\right) = \underline{h}\left(\underline{g_1}\left(\underline{\vec{m}}\right), \ldots, \underline{g_l}\left(\underline{\vec{m}}\right)\right).$$

5. Let $f = \mathsf{Rec}(h, g)$ for some n-ary primitive recursive function h and some $(n+2)$-ary primitive recursive function g. Here we use $\vec{m}'$ for $m_1, \ldots, m_{n-1}$ and $\underline{\vec{m}}'$ for $\underline{m_1}, \ldots, \underline{m_{n-1}}$ to denote the (possibly empty) list resulting from the omission of m_0, respectively $\underline{m_0}$. And the omitted m_0 is used for subinduction.

If $m_0 = 0$, then by the outer induction hypothesis for h,

$$\mathsf{PA} \vdash \underline{h}\left(\underline{\vec{m}}'\right) = \underline{h(\vec{m}')}. \quad (6.6)$$

Since $[\mathsf{Rec}(h, g)](0, \vec{m}') = h(\vec{m}')$, we need to show that

$$\mathsf{PA} \vdash \underline{\mathsf{Rec}(h, g)}\left(0, \underline{\vec{m}}'\right) = \underline{h(\vec{m}')},$$

which follows by first-order reasoning from (6.6), and the axioms for primitive recursion that guarantee

$$\mathsf{PA} \vdash \underline{\mathsf{Rec}(h, g)}\left(0, \underline{\vec{m}}'\right) = \underline{h}\left(\underline{\vec{m}}'\right).$$

6.2 Preliminaries

If $m_0 > 0$, then by the axioms for the successor function,

$$\mathsf{PA} \vdash \underline{m_0} = \mathsf{s}\left(\underline{m_0 - 1}\right) \tag{6.7}$$

and by the outer induction hypothesis for g,

$$\mathsf{PA} \vdash \underline{g\big([\mathsf{Rec}(h,g)](m_0-1,\vec{m}'),\ m_0-1,\ \vec{m}'\big)} = \underline{g\big([\mathsf{Rec}(h,g)](m_0-1,\vec{m}'),\ m_0-1,\ \vec{m}'\big)}. \tag{6.8}$$

Further, by the subinduction hypothesis for $m_0 - 1$,

$$\mathsf{PA} \vdash \underline{\mathsf{Rec}(h,g)}\left(\underline{m_0-1},\underline{\vec{m}'}\right) = \underline{[\mathsf{Rec}(h,g)](m_0-1,\vec{m}')}. \tag{6.9}$$

Since

$$[\mathsf{Rec}(h,g)](m_0,\vec{m}') = g\big([\mathsf{Rec}(h,g)](m_0-1,\vec{m}'),\ m_0-1,\ \vec{m}'\big),$$

we need to show

$$\mathsf{PA} \vdash \underline{\mathsf{Rec}(h,g)}\left(\underline{m_0},\underline{\vec{m}'}\right) = \underline{g\big([\mathsf{Rec}(h,g)](m_0-1,\vec{m}'),\ m_0-1,\ \vec{m}'\big)},$$

which follows by first-order reasoning from (6.7), (6.8), (6.9), and the axioms for primitive recursion that guarantee that

$$\mathsf{PA} \vdash \underline{\mathsf{Rec}(h,g)}\left(\mathsf{s}\left(\underline{m_0-1}\right),\underline{\vec{m}'}\right) = \underline{g\left(\mathsf{Rec}(h,g)\left(\underline{m_0-1},\underline{\vec{m}'}\right),\ \underline{m_0-1},\ \underline{\vec{m}'}\right)}.$$

$\square$

Another property we will need states that each term of $\mathcal{L}_{\mathsf{PA}}$ provably corresponds to some primitive recursive function[12] and that each ground term of $\mathcal{L}_{\mathsf{PA}}$ provably equals its value in the standard model:

---

[12] In fact, there are infinitely many primitive recursive functions differing in their arity and in the order of arguments. The latter difference is due to the fact that the order of free variables in a formula is not fixed but the order of arguments of a primitive recursive function is.

Lemma 6.18. *Let t be a term of $\mathcal{L}_{\mathsf{PA}}$ and let all variables occurring (freely) in t be present in the sequence $\vec{x} = x_0, \ldots, x_{n-1}$ of variables (without repetitions). Then there exists an n-ary primitive recursive function f such that*
$$\mathsf{PA} \vdash t = \underline{f}(\vec{x}) \ .$$
Moreover, for each ground term t, we have
$$\mathsf{PA} \vdash t = \underline{t^{\mathbb{N}}} \ .$$

Proof. We proceed by induction on the structure of the term t. We distinguish the following cases:

1. If $t = y$, then $y = x_i$ for some $0 \le i < n$. Note that y is not a ground term. For $f := \mathsf{pr}_i^n$ by the axioms for the projection functions $\mathsf{PA} \vdash y = \underline{\mathsf{pr}_i^n}(\vec{x})$.

2. If $t = \underline{m}$, for $f := \mathsf{cst}_m^n$ by the axioms for the constant functions $\mathsf{PA} \vdash \underline{m} = \underline{\mathsf{cst}_m^n}(\vec{x})$. Moreover, since $\underline{m}^{\mathbb{N}} = m$, we have $\underline{t^{\mathbb{N}}} = \underline{m}$, making $\mathsf{PA} \vdash t = \underline{t^{\mathbb{N}}}$ a trivial consequence of the equality axioms.

3. The remaining case is $t = \underline{h}(t_1, \ldots, t_m)$ for some terms $t_1, \ldots, t_m$ and some m-ary primitive recursive function h with $m \ge 1$. Since all variables occurring in any of t_i's also occur in t, we can use the induction hypothesis for the same n, meaning that for each $1 \le i \le m$ there exists an n-ary primitive recursive function g_i such that $\mathsf{PA} \vdash t_i = \underline{g_i}(\vec{x})$. Setting $f := \mathsf{Comp}(h, g_1, \ldots, g_m)$, we need to show that
$$\mathsf{PA} \vdash t = \underline{\mathsf{Comp}(h, g_1, \ldots, g_m)}(\vec{x}) \ ,$$
which follows from the induction hypotheses above and the axioms for function composition
$$\mathsf{PA} \vdash \underline{\mathsf{Comp}(h, g_1, \ldots, g_m)}(\vec{x}) = \underline{h}\left(\underline{g_1}(\vec{x}), \ldots, \underline{g_m}(\vec{x})\right) \ .$$
In addition, if t is a ground term, so are the terms $t_1, \ldots, t_m$. Thus, by the induction hypothesis for ground terms t_i, we have $\mathsf{PA} \vdash t_i = \underline{t_i^{\mathbb{N}}}$ for each $1 \le i \le m$. Given that $t^{\mathbb{N}} = h(t_1^{\mathbb{N}}, \ldots, t_m^{\mathbb{N}})$, we need to show that
$$\mathsf{PA} \vdash \underline{h}(t_1, \ldots, t_m) = \underline{h}(\underline{t_1^{\mathbb{N}}}, \ldots, \underline{t_m^{\mathbb{N}}}) \ ,$$
which follows from the induction hypotheses above by Lemma 6.17.
□

6.2 Preliminaries

Definition 6.19 (Provable equivalence). Let ϕ and ψ be $\mathcal{L}_{\mathsf{PA}}$-formulas. We say that ϕ and ψ are *provably equivalent* if
$$\mathsf{PA} \vdash \phi \leftrightarrow \psi \ .$$

Now we can define several important classes of $\mathcal{L}_{\mathsf{PA}}$-formulas.

Definition 6.20 (Standard primitive recursive, standard Σ_1-, provably Σ_1, and provably Δ_1 formulas).

1. A *standard primitive recursive formula* is an $\mathcal{L}_{\mathsf{PA}}$-formula of the form
$$\underline{R}(t_0, \ldots, t_{n-1})$$
where R is an n-ary primitive recursive relation for some $n \geq 1$.

2. A *standard Σ_1-formula* is an $\mathcal{L}_{\mathsf{PA}}$-formula of the form
$$\exists x \phi$$
where ϕ is a standard primitive recursive formula.

3. An $\mathcal{L}_{\mathsf{PA}}$-formula ϕ is *provably Σ_1* if there exists a standard Σ_1-formula ψ such that ϕ and ψ are provably equivalent.

4. An $\mathcal{L}_{\mathsf{PA}}$-formula ϕ is *provably Δ_1* if both ϕ and $\neg\phi$ are provably Σ_1.

Lemma 6.21. *Provably Δ_1 formulas are closed under Boolean connectives and under substitutions of terms for variables.*

Proof. We first show the closure under Boolean connectives.

We have that $\bot$ is provably equivalent to $\exists x 0 = 1$ and $\neg\bot$ is provably equivalent to $\exists x 0 = 0$. It follows that $\bot$ is provably Δ_1.

Further, let ϕ and ψ be provably Δ_1 formulas such that

$$\mathsf{PA} \vdash \phi \leftrightarrow \exists x \phi', \qquad \mathsf{PA} \vdash \psi \leftrightarrow \exists x \psi',$$
$$\mathsf{PA} \vdash \neg\phi \leftrightarrow \exists x \phi'', \qquad \mathsf{PA} \vdash \neg\psi \leftrightarrow \exists x \psi''$$

for some standard primitive recursive formulas ϕ', ϕ'', ψ', and ψ''. We find that $\phi \to \psi$ is provably equivalent to $\neg\phi \lor \psi$, which is provably equivalent to $\exists x \phi'' \lor \exists x \psi'$, which is provably equivalent to $\exists x (\phi'' \lor \psi')$. Thus, it remains to show that $\phi'' \lor \psi'$ is provably equivalent to a standard primitive recursive formula. We know that standard primitive recursive formulas ϕ'' and ψ' must have the form $\underline{R}(t_0, \ldots, t_{m-1})$ and

$Q(s_0, \ldots, s_{k-1})$ for some terms $t_0, \ldots, t_{m-1}, s_0, \ldots, s_{k-1}$ and some primitive recursive relations R and Q that are m-ary and k-ary respectively. Let $\vec{z} = z_0, \ldots, z_{n-1}$ be a list of all variables occurring (freely) in ϕ'' and ψ', taken in a particular but arbitrary order. By the axioms for primitive recursive relations,

$$\mathsf{PA} \vdash \underline{R(t_0, \ldots, t_{m-1})} \leftrightarrow \underline{\chi_R(t_0, \ldots, t_{m-1}) = 0} \;,$$
$$\mathsf{PA} \vdash \underline{Q(s_0, \ldots, s_{k-1})} \leftrightarrow \underline{\chi_Q(s_0, \ldots, s_{k-1}) = 0} \;.$$

By Lemma 6.18 applied to the terms

$$\underline{\chi_R(t_0, \ldots, t_{m-1})} \quad \text{and} \quad \underline{\chi_Q(s_0, \ldots, s_{k-1})} \;,$$

there exist n-ary primitive recursive functions f and g such that

$$\mathsf{PA} \vdash \underline{\chi_R(t_0, \ldots, t_{m-1})} = \underline{f(\vec{z})} \;,$$
$$\mathsf{PA} \vdash \underline{\chi_Q(s_0, \ldots, s_{k-1})} = \underline{g(\vec{z})} \;.$$

Thus,

$$\mathsf{PA} \vdash \phi'' \leftrightarrow \underline{f(\vec{z}) = 0} \;,$$
$$\mathsf{PA} \vdash \psi' \leftrightarrow \underline{g(\vec{z}) = 0} \;.$$

For the n-ary primitive recursive function

$$h := \mathsf{Comp}(\cdot, f, g)$$

we find by the axioms for function composition

$$\mathsf{PA} \vdash \underline{h(\vec{z})} = \underline{f(\vec{z})} \cdot \underline{g(\vec{z})} \;.$$

Note that $y_1 \cdot y_2 = 0 \leftrightarrow y_1 = 0 \lor y_2 = 0$ is a formula both of our formulation of PA and of the weaker formulation PA' from [109]. As follows from

6.2 Preliminaries

[109, Lemma 3.1(S7′),Proposition 3.2(l,n), and Proposition 3.5(e)][13],

$$\mathsf{PA}' \vdash y_1 \cdot y_2 = 0 \leftrightarrow y_1 = 0 \lor y_2 = 0 \ .$$

Therefore, by Remark 6.16 and first-order reasoning

$$\mathsf{PA} \vdash \underline{h}(\vec{z}) = 0 \leftrightarrow \underline{f}(\vec{z}) = 0 \lor \underline{g}(\vec{z}) = 0 \ .$$

Accordingly,

$$\mathsf{PA} \vdash \phi'' \lor \psi' \leftrightarrow \underline{h}(\vec{z}) = 0 \ .$$

It remains to show that h is a characteristic function (of some primitive recursive relation). It is easy to see that the only values that the functions f and g can take are 0 and 1. Let us show this for the function f; the argument for g is identical. Let $f(\vec{k}) = l$ for some natural numbers $\vec{k} = k_0, \ldots, k_{n-1}$ and a natural number l. Then by Lemma 6.17, for $\vec{k} = \underline{k_0}, \ldots, \underline{k_{n-1}}$, we have $\mathsf{PA} \vdash \underline{f}(\vec{k}) = \underline{l}$. It immediately follows that

$$\mathsf{PA} \vdash \left(\underline{\chi_R}(t_0, \ldots, t_{m-1})\right) \left[\vec{z} \mapsto \vec{\underline{k}}\right] = \underline{l} \ .$$

In other words,

$$\mathsf{PA} \vdash \underline{\chi_R} \left(t_0 \left[\vec{z} \mapsto \vec{\underline{k}}\right], \ldots, t_{m-1} \left[\vec{z} \mapsto \vec{\underline{k}}\right]\right) = \underline{l} \ .$$

Since $t_i[\vec{z} \mapsto \vec{\underline{k}}]$ are ground terms, for $j_i = \left(t_i[\vec{z} \mapsto \vec{\underline{k}}]\right)^{\mathbb{N}}$ for

$$i = 0, \ldots, m-1$$

we have by Lemma 6.18 that

$$\mathsf{PA} \vdash \underline{\chi_R}\left(\underline{j_0}, \ldots, \underline{j_{m-1}}\right) = \underline{l} \ .$$

---

[13] The number of statements (which are supplied with complete proofs in [109]) necessary to demonstrate this simple arithmetic fact suggests the level of tediousness involved. These proofs with minimal change can also be carried out in our PA. However, as our book is not a book on formal arithmetic, we choose to omit these proofs and similar proofs of well-known properties of natural numbers. From now on we will omit even the reference to Remark 6.16. An interested reader can restore the proofs with the help of any standard textbook on Peano Arithmetic, e.g., [109]. The only thing to keep in mind is that the formulation of PA′ in [109] includes axioms about + and ·. It would seem that these axioms are lacking in our formulation. However, this impression is wrong. These axioms are present among other defining axioms for primitive recursive functions. To see that, it is sufficient to recall that multiplication can be obtained by the scheme of primitive recursion from addition, and similarly addition from the successor function.

Since χ_R is a characteristic function, it follows by Lemma 6.17 that either $\mathsf{PA} \vdash \underline{l} = \underline{0}$ or $\mathsf{PA} \vdash \underline{l} = \underline{1}$. It now remains to invoke the soundness of PA, (6.3), to conclude that $l = 0$ or $l = 1$. Just like f and g, the n-ary primitive recursive function h can only take values 0 and 1. Accordingly, h is the characteristic function of some n-ary primitive recursive relation S such that
$$\mathsf{PA} \vdash \underline{S}(\vec{z}) \leftrightarrow \underline{h}(\vec{z}) = \underline{0} \ .$$
Thus, $\phi'' \vee \psi'$ is provably equivalent to the standard primitive recursive formula $\underline{S}(\vec{z})$. This concludes the proof that $\phi \to \psi$ is provably Σ_1.

We now show that $\neg(\phi \to \psi)$ is also provably Σ_1. We know that it is provably equivalent to $\exists x \phi' \wedge \exists x \psi''$, which is provably equivalent to
$$\exists y \Big(y = \langle (y)_0, (y)_1 \rangle \ \wedge \ \phi'[x \mapsto (y)_0] \ \wedge \ \psi''[x \mapsto (y)_1] \Big) \ .$$
Since ϕ' and ψ'' are standard primitive recursive formulas, so are their substitution instances $\phi'[x \mapsto (y)_0]$ and $\psi''[x \mapsto (y)_1]$. Note further that $\langle (y)_0, (y)_1 \rangle$ is a term and $=$ is a binary relation symbol, making also $y = \langle (y)_0, (y)_1 \rangle$ a standard primitive recursive formula. Similar to the case of $\phi'' \to \psi'$, there exist $(n+1)$-ary primitive recursive characteristic functions f, g_1, and g_2 such that
$$\mathsf{PA} \vdash y = \langle (y)_0, (y)_1 \rangle \leftrightarrow \underline{f}(y, \vec{z}) = \underline{0} \ ,$$
$$\mathsf{PA} \vdash \phi'[x \mapsto (y)_0] \leftrightarrow \underline{g_1}(y, \vec{z}) = \underline{0} \ ,$$
$$\mathsf{PA} \vdash \psi''[x \mapsto (y)_1] \leftrightarrow \underline{g_2}(y, \vec{z}) = \underline{0} \ ,$$
where $\vec{z} = z_0, \ldots, z_{n-1}$ is a sequence (without repetitions) of all variables that occur (freely) in ϕ' and ψ'' except for x. Then the primitive recursive function
$$h(l, \vec{k}) := 1 \dotminus \Big((1 \dotminus f(l, \vec{k})) \cdot (1 \dotminus g_1(l, \vec{k})) \cdot (1 \dotminus g_2(l, \vec{k})) \Big) \ ,$$
where 1 stands for the function $\mathsf{cst}_1^{n+1}(l, \vec{k})$, is a characteristic function of some $(n+1)$-ary primitive recursive relation S such that $\neg(\phi \to \psi)$ is provably equivalent to $\exists y \underline{S}(y, \vec{z})$.

To establish closure under substitutions, we show that provably Σ_1 formulas satisfy this closure property. Indeed, let $\phi(y)$ be provably Σ_1, that is
$$\mathsf{PA} \vdash \phi(y) \leftrightarrow \exists x \psi(y)$$

6.2 Preliminaries

for some standard primitive recursive formula $\psi(y)$. Thus $\psi(t)$ also is a standard primitive recursive formula and

$$\mathsf{PA} \vdash \phi(t) \leftrightarrow \exists x \psi(t) \ .$$

Hence $\phi(t)$ is provably Σ_1. Now it immediately follows that provably Δ_1 formulas are closed under substitutions of terms for variables. □

Now we can establish the Σ_1-completeness of PA.

Lemma 6.22 (Σ_1-completeness). *For any provably Σ_1 sentence ϕ,*

$$\mathbb{N} \vDash \phi \quad \Longrightarrow \quad \mathsf{PA} \vdash \phi \ .$$

Proof. We start by proving completeness for standard primitive recursive sentences. Let $\mathbb{N} \vDash \phi$ for a standard primitive recursive sentence ϕ. Since ϕ has the form $\underline{R}(t_0, \ldots, t_{n-1})$, where $n \geq 1$, $t_0, \ldots, t_{n-1}$ are ground terms, and R is an n-ary primitive recursive relation, we find by the definition of $\mathbb{N} \vDash \phi$ that

$$\chi_R(t_0^{\mathbb{N}}, \ldots, t_{n-1}^{\mathbb{N}}) = 0 \ .$$

By the definition of $\cdot^{\mathbb{N}}$ we find

$$(\underline{\chi_R}(t_0, \ldots, t_{n-1}))^{\mathbb{N}} = \chi_R(t_0^{\mathbb{N}}, \ldots, t_{n-1}^{\mathbb{N}}) = 0 \ . \quad (6.10)$$

By Lemma 6.18 for the ground term $\underline{\chi_R}(t_0, \ldots, t_{n-1})$ we obtain

$$\mathsf{PA} \vdash \underline{\chi_R}(t_0, \ldots, t_{n-1}) = \underline{(\underline{\chi_R}(t_0, \ldots, t_{n-1}))^{\mathbb{N}}} \ .$$

By (6.10) this is

$$\mathsf{PA} \vdash \underline{\chi_R}(t_0, \ldots, t_{n-1}) = 0 \ .$$

Finally, using the axioms for primitive recursive relations yields

$$\mathsf{PA} \vdash \underline{R}(t_0, \ldots, t_{n-1}) \ ,$$

i.e., $\mathsf{PA} \vdash \phi$.

Let us now extend the completeness result to standard Σ_1-sentences. Let $\mathbb{N} \vDash \phi$ for a standard Σ_1-sentence ϕ, which has the form $\exists x \psi(x)$ for some standard primitive recursive formula $\psi(x)$ that has no free occurrences of variables other than x. It follows that $\mathbb{N} \vDash \psi(\underline{n})$ for some natural number n. By the already proved completeness for standard primitive recursive sentences, we have $\mathsf{PA} \vdash \psi(\underline{n})$ for the standard primitive recursive sentence $\psi(\underline{n})$ and, thus, by first-order reasoning, $\mathsf{PA} \vdash \exists x \psi(x)$, i.e., $\mathsf{PA} \vdash \phi$.

The remaining step to provably Σ_1 sentences is now simple. Let $\mathbb{N} \vDash \phi$ for a provably Σ_1 sentence ϕ. Then there is a standard Σ_1-formula ψ such that $\mathsf{PA} \vdash \phi \leftrightarrow \psi$. In order to use completeness for standard Σ_1-sentences, however, we would need ψ to be a sentence. Let ψ' be obtained from ψ by simultaneously substituting 0 for each variable occurring free in ψ. Since the same substitution does not affect the sentence ϕ, it follows by first-order reasoning that $\mathsf{PA} \vdash \phi \leftrightarrow \psi'$. Further, it is easy to see that ψ' is a standard Σ_1-sentence. Since $\mathbb{N} \vDash \phi \leftrightarrow \psi'$ by the soundness of PA, it follows that $\mathbb{N} \vDash \psi'$. By the already proved completeness for standard Σ_1-sentences, we have $\mathsf{PA} \vdash \psi'$, which implies $\mathsf{PA} \vdash \phi$. □

Applying Lemma 6.22 to provably Δ_1 sentences yields the following theorem, which we are going to apply often:

Theorem 6.23 (Δ_1-completeness). *For any provably Δ_1 sentence ϕ,*

1. *if $\mathbb{N} \vDash \phi$, then $\mathsf{PA} \vdash \phi$;*

2. *if $\mathbb{N} \nvDash \phi$, then $\mathsf{PA} \vdash \neg \phi$.*

Proof. Since any provably Δ_1 formula is also provably Σ_1, the first claim is an instance of Σ_1-completeness.

To show the second claim, assume that $\mathbb{N} \nvDash \phi$ for a provably Δ_1 sentence ϕ. Then $\neg \phi$ is a provably Σ_1 sentence. From $\mathbb{N} \vDash \neg \phi$ we obtain $\mathsf{PA} \vdash \neg \phi$ by Σ_1-completeness. □

To be able to talk about formulas and proofs of PA within PA, we need a so-called Gödel numbering of $\mathcal{L}_{\mathsf{PA}}$. In later sections, we also use a Gödel numbering for formulas and terms of $\mathcal{L}_\mathsf{J}$, as well as for schemes of formulas of $\mathcal{L}_\mathsf{J}$. Thus, from the very beginning we use a joint injective Gödel numbering for both languages with the language of $\mathcal{L}_\mathsf{J}$ including four types of variables: atomic propositions, variables over formulas (for schemes), justification variables, and variables over terms (for schemes). That is an assignment of a numerical code $\ulcorner \phi \urcorner$ to each formula $\phi \in \mathcal{L}_{\mathsf{PA}}$, of a numerical code $\ulcorner A \urcorner$ to each formula $A \in \mathcal{L}_\mathsf{J}$, and of numerical codes $\ulcorner r \urcorner$ and $\ulcorner t \urcorner$ to each arithmetical term r and to each justification term t. The injectivity requirement means that Gödel numbers of distinct objects must be different, i.e., $\ulcorner X \urcorner = \ulcorner Y \urcorner$ implies

6.2 Preliminaries

$X = Y$ for arbitrary syntactic objects X and Y. Moreover, we require that $\langle \ulcorner X_0 \urcorner, \ldots, \ulcorner X_{n-1} \urcorner \rangle \neq \ulcorner Y \urcorner$ for any $n \geq 0$ and any syntactic objects $X_0, \ldots, X_{n-1}$ and Y.

As usual, we assume that simple syntactic operations on arithmetical formulas correspond to primitive recursive functions on their Gödel numbers. In particular, we will use the following functions

- subst is a ternary primitive recursive function such that, for all formulas $\phi \in \mathcal{L}_{\mathsf{PA}}$, all arithmetical variables y, and all arithmetical terms t,
$$\mathsf{subst}(\ulcorner \phi \urcorner, \ulcorner t \urcorner, \ulcorner y \urcorner) = \ulcorner \phi[y \mapsto t] \urcorner \ . \tag{6.11}$$

- num is the unary primitive recursive function such that, for all natural numbers k,
$$\mathsf{num}(k) = \ulcorner \underline{\mathsf{cst}_k^0} \urcorner = \ulcorner \underline{k} \urcorner \ . \tag{6.12}$$

- not is a unary primitive recursive function such that, for all formulas $\phi \in \mathcal{L}_{\mathsf{PA}}$,
$$\mathsf{not}(\ulcorner \phi \urcorner) = \ulcorner \neg \phi \urcorner \ . \tag{6.13}$$

- trank is a unary primitive recursive function such that, for all justification terms $t \in \mathcal{L}_{\mathsf{J}}$,
$$\mathsf{trank}(\ulcorner t \urcorner) = \mathsf{rk}(t) \tag{6.14}$$

  where rk is the rank function given in Definition 4.28.

As before, we choose the functions subst, not and trank with the only requirement for its values for inputs not of the specified format being that the resulting function be primitive recursive.

Making use of the Gödel numbering, we can state the Diagonalization Lemma, which is crucial for defining arithmetical interpretations for $\mathsf{LP}_{\mathsf{CS}}$.

Lemma 6.24 (Diagonalization). *Let $\psi(y)$ be an $\mathcal{L}_{\mathsf{PA}}$-formula. There exists an $\mathcal{L}_{\mathsf{PA}}$-formula ϕ such that*

- $\mathsf{PA} \vdash \phi \leftrightarrow \psi(\ulcorner \phi \urcorner)$;

- *a variable $x \neq y$ has a free occurrence in ϕ if and only if it occurs in ψ; the variable y does not occur freely in ϕ;*

- if ψ is provably Δ_1, then so is ϕ; and
- if ψ contains a relation symbol $\underline{R}$, then so does ϕ.

Proof. Let f be a unary primitive recursive function such that
$$f(l) = \mathsf{subst}(l, \mathsf{num}(l), \ulcorner y \urcorner)$$
for each natural number l. We set
$$\phi'(y) := \psi\left(\underline{f(y)}\right) \ .$$
Further, let
$$k := \ulcorner \phi'(y) \urcorner \ .$$
By Lemma 6.17, we have
$$\mathsf{PA} \vdash \underline{f(\underline{k})} = \ulcorner \phi'(\underline{k}) \urcorner \tag{6.15}$$
because, by (6.11) and (6.12),
$$f(k) = \mathsf{subst}(k, \mathsf{num}(k), \ulcorner y \urcorner) = \mathsf{subst}(\ulcorner \phi'(y) \urcorner, \ulcorner \underline{k} \urcorner, \ulcorner y \urcorner) = \ulcorner \phi'(\underline{k}) \urcorner \ .$$
Now we define
$$\phi := \phi'(\underline{k}) \ ,$$
which yields, by the definition of ϕ', that
$$\phi = \psi\left(\underline{f(k)}\right) \ . \tag{6.16}$$
Moreover, by (6.15) we have
$$\mathsf{PA} \vdash \underline{f(\underline{k})} = \ulcorner \phi \urcorner \ .$$
Therefore, based on the equality axioms we conclude
$$\mathsf{PA} \vdash \phi \leftrightarrow \psi(\ulcorner \phi \urcorner) \ .$$

By (6.16), we see that ϕ is obtained from ψ by substituting a ground term for the variable y. Thus,

1. the free variables of ϕ are the same as in ψ, except that y is not free in ϕ;
2. by Lemma 6.21, ϕ is provably Δ_1 whenever ψ is, and

6.2 Preliminaries

3. ϕ contains exactly the same relation symbols as ψ. □

Remark 6.25. Compared to the standard formulation of the Diagonalization Lemma, we have added the requirement that ϕ contain all the relation symbols that ψ does. This additional requirement does not require any modification to the construction of ϕ and is to be used for proving arithmetical completeness of the Logic of Proofs.

Definition 6.26 (Proof predicate). A *proof predicate for a theory* T is a provably Δ_1 formula $\mathsf{Prf}_\mathsf{T}(x,y)$ with no free variables other than x and y such that for every $\mathcal{L}_{\mathsf{PA}}$-sentence ϕ,

$$\mathsf{T} \vdash \phi \quad \text{if and only if} \\ \mathbb{N} \vDash \mathsf{Prf}_\mathsf{T}(\underline{n}, \ulcorner \phi \urcorner) \text{ for some natural number } n. \quad (6.17)$$

We will not formally establish the existence of proof predicates for PA. Later, in Section 6.4, we present an informal construction of a particular proof predicate that is needed to obtain the arithmetical interpretation for the Logic of Proofs. A detailed formal construction of a proof predicate is presented, e.g., in [69]. For the rest of this chapter, we simply assume that we are given a proof predicate for PA, which we denote by $\mathsf{Proof}(x,y)$.

Although we will not need this later, we would like to finish this section with showing Gödel–Rosser Incompleteness Theorem. For more details, we refer again to [69].

Theorem 6.27 (Gödel–Rosser). *Let* T *be a theory in* $\mathcal{L}_{\mathsf{PA}}$ *that contains* PA. *Assume that there exists a proof predicate* $\mathsf{Prf}_\mathsf{T}(x,y)$ *for* T. *If* T *is consistent, there is an* $\mathcal{L}_{\mathsf{PA}}$-*sentence* ϕ *with* $\mathbb{N} \vDash \phi$ *such that* $\mathsf{T} \nvdash \phi$ *and* $\mathsf{T} \nvdash \neg \phi$.

Proof. Set $\mathsf{Ref}(x,y) := \mathsf{Prf}_\mathsf{T}(x, \underline{\mathsf{not}}(y))$. Since $\mathsf{Prf}_\mathsf{T}(x,y)$ is provably Δ_1 by the definition of proof predicates, so is $\mathsf{Ref}(x,y)$ by Lemma 6.21. We define

$$\psi(u) := \neg \exists x \Big(\mathsf{Prf}_\mathsf{T}(x,u) \wedge (\forall y < x) \neg \mathsf{Ref}(y, u) \Big) \ .$$

By the Diagonalization Lemma 6.24 applied to $\psi(u)$, there is a formula ϕ such that

$$\mathsf{PA} \vdash \phi \leftrightarrow \neg \exists x \Big(\mathsf{Prf}_\mathsf{T}(x, \ulcorner \phi \urcorner) \wedge (\forall y < x) \neg \mathsf{Ref}(y, \ulcorner \phi \urcorner) \Big) \ . \quad (6.18)$$

Since u is the only free variable of $\psi(u)$, the formula ϕ is actually a sentence.

Suppose towards a contradiction that $\mathsf{T} \vdash \phi$. Then, by (6.17), there would exist a natural number m such that
$$\mathbb{N} \models \mathsf{Prf}_\mathsf{T}(\underline{m}, \ulcorner \phi \urcorner) \ .$$

Since, by assumption, T is consistent, we would have $\mathsf{T} \nvdash \neg \phi$. Thus, by (6.17), (6.13), and Lemma 6.17, we would have
$$\mathbb{N} \nvDash \mathsf{Ref}(\underline{n}, \ulcorner \phi \urcorner) \quad \text{for any natural number } n.$$

Taking into account that both $\mathsf{Prf}_\mathsf{T}(\underline{m}, \ulcorner \phi \urcorner)$ and $\mathsf{Ref}(\underline{n}, \ulcorner \phi \urcorner)$ are provably Δ_1 sentences by Lemma 6.21, by Δ_1-completeness (Theorem 6.23), we would have
$$\mathsf{PA} \vdash \mathsf{Prf}_\mathsf{T}(\underline{m}, \ulcorner \phi \urcorner) \quad \text{and} \quad \mathsf{PA} \vdash \neg \mathsf{Ref}(\underline{n}, \ulcorner \phi \urcorner) \quad \text{for all } n.$$

The same argument as used for [109, Proposition 3.8(b′)] shows that, for any formula $A(x)$,
$$\mathsf{PA} \ \vdash \ A(\underline{0}) \wedge A(\underline{1}) \wedge \cdots \wedge A(\underline{m-1}) \ \leftrightarrow \ (\forall x < \underline{m}) A(x) \ .$$

Hence, we would have
$$\mathsf{PA} \vdash \mathsf{Prf}_\mathsf{T}(\underline{m}, \ulcorner \phi \urcorner) \wedge (\forall y < \underline{m}) \neg \mathsf{Ref}(y, \ulcorner \phi \urcorner) \ ,$$

which would yield
$$\mathsf{PA} \vdash \exists x \Big(\mathsf{Prf}_\mathsf{T}(x, \ulcorner \phi \urcorner) \wedge (\forall y < x) \neg \mathsf{Ref}(y, \ulcorner \phi \urcorner) \Big) \ .$$

By (6.18), we would get $\mathsf{PA} \vdash \neg \phi$ and since PA is contained in T we would also get $\mathsf{T} \vdash \neg \phi$, which, for a consistent theory T, would contradict the initial assumption $\mathsf{T} \vdash \phi$.

Next we show that $\mathsf{T} \nvdash \neg \phi$. Suppose towards a contradiction that $\mathsf{T} \vdash \neg \phi$. In this case, by (6.17), (6.13), and Lemma 6.17, there would exist a natural number m such that
$$\mathbb{N} \models \mathsf{Ref}(\underline{m}, \ulcorner \phi \urcorner) \ .$$

Since T is assumed to be consistent, we would have $\mathsf{T} \nvdash \phi$, meaning that
$$\mathbb{N} \nvDash \mathsf{Prf}_\mathsf{T}(\underline{n}, \ulcorner \phi \urcorner) \quad \text{for any natural number } n$$

6.2 Preliminaries

by (6.17). By similar arguments as before, from this we would obtain

$$\mathsf{PA} \vdash \mathsf{Ref}(\underline{m}, \ulcorner \phi \urcorner) \quad \text{and} \quad \mathsf{PA} \vdash \neg \mathsf{Prf}_\mathsf{T}(\underline{n}, \ulcorner \phi \urcorner) \quad \text{for all } n \ .$$

We conclude that the following would hold:

$$\mathsf{PA} \vdash \forall x \Big(\big(x \leq \underline{m} \to \neg \mathsf{Prf}_\mathsf{T}(x, \ulcorner \phi \urcorner) \big) \land \big(\underline{m} < x \to (\exists y < x) \mathsf{Ref}(y, \ulcorner \phi \urcorner) \big) \Big) \ .$$

Therefore, by $\mathsf{PA} \vdash \forall x (x \leq \underline{m} \lor \underline{m} < x)$ we would get

$$\mathsf{PA} \vdash \forall x \Big(\neg \mathsf{Prf}_\mathsf{T}(x, \ulcorner \phi \urcorner) \lor (\exists y < x) \mathsf{Ref}(y, \ulcorner \phi \urcorner) \Big) \ .$$

That is, equivalently, we would have

$$\mathsf{PA} \vdash \neg \exists x \Big(\mathsf{Prf}_\mathsf{T}(x, \ulcorner \phi \urcorner) \land (\forall y < x) \neg \mathsf{Ref}(y, \ulcorner \phi \urcorner) \Big) \ .$$

Hence, by (6.18), we would get $\mathsf{PA} \vdash \phi$. Again this would imply $\mathsf{T} \vdash \phi$ and since T is assumed to be consistent, we would get the desired contradiction.

It remains to show that $\mathbb{N} \models \phi$. First, we observe that by the soundness of PA we get from (6.18) that

$$\mathbb{N} \models \phi \leftrightarrow \neg \exists x \Big(\mathsf{Prf}_\mathsf{T}(x, \ulcorner \phi \urcorner) \land (\forall y < x) \neg \mathsf{Ref}(y, \ulcorner \phi \urcorner) \Big) \ . \tag{6.19}$$

From $\mathsf{T} \nvdash \phi$, it follows by (6.17) that $\mathbb{N} \nvDash \mathsf{Prf}_\mathsf{T}(\underline{n}, \ulcorner \phi \urcorner)$ for any natural number n. We immediately obtain

$$\mathbb{N} \models \neg \exists x \Big(\mathsf{Prf}_\mathsf{T}(x, \ulcorner \phi \urcorner) \land (\forall y < x) \neg \mathsf{Ref}(y, \ulcorner \phi \urcorner) \Big)$$

and hence by (6.19) we conclude that $\mathbb{N} \models \phi$. □

We obtain Gödel's First Incompleteness Theorem from Theorem 6.27 by replacing the assumption that T is consistent with the stronger assumption that T is ω-consistent. A theory T is called ω-*consistent* if for all $\mathcal{L}_\mathsf{PA}$-formulas $\phi(x)$ we have

if $\mathsf{T} \vdash \phi(\underline{n})$ for all natural numbers n, then $\mathsf{T} + \forall x \phi(x)$ is consistent.

Gödel proved his famous result in [71]. The improved version of of Theorem 6.27 is due to Rosser [121]. He contributed the extra conjunct

$$(\forall y < x) \neg \mathsf{Ref}(y, \ulcorner \phi \urcorner)$$

in the definition of $\psi(u)$, thereby obviating ω-consistency.

6.3 Simple Arithmetical Interpretation

The methods we develop for obtaining arithmetical interpretations of $\mathsf{LP}_{\mathsf{CS}}$ are based on our decidability results for the Logic of Proofs. In particular, we make crucial use of finitary models for $\mathsf{LP}_{\mathsf{CS}}$. We start with recalling those results that are essential for the purpose of studying arithmetical interpretations.

First we have completeness of $\mathsf{LP}_{\mathsf{CS}}$ with respect to finitary models, which we showed in Theorem 4.35.

Theorem 4.35. *Let CS be an almost schematic constant specification. Let F be a formula that is not derivable in $\mathsf{LP}_{\mathsf{CS}}$. Then there exists a finitary CS-model $\mathcal{M}_{\mathit{fin}}$ such that $\mathcal{M}_{\mathit{fin}} \not\models F$.*

In Lemma 4.30 and Theorem 4.31, we showed that the evidence function and the satisfaction relation for finitary models is decidable. In the context of arithmetical interpretations, we need a stronger result: namely, that the evidence function and the satisfaction relation are primitive recursive. However, primitive recursiveness is a property of natural numbers, meaning that Lemma 4.30 and Theorem 4.31 have to be reformulated using Gödel numbering. In addition, we need a coding of finitary models and their ingredients into natural numbers. Recall also that our Gödel numbering covers the syntax of $\mathcal{L}_\mathsf{J}$ in addition to that of $\mathcal{L}_\mathsf{PA}$.

Definition 6.28 (Primitive recursive constant specification). A constant specification CS is called *primitive recursive* if the binary relation (on natural numbers)

$$\mathsf{CS}_\mathbb{N} := \{(\ulcorner c \urcorner, \ulcorner A \urcorner) \mid (c, A) \in \mathsf{CS}\}$$

is primitive recursive.

Since showing that a given function is primitive recursive by precise coding is quite tedious (and has been done elsewhere multiple times), we will use the following high-level criterion that can be found, e.g., in [124, Theorem 75]:

Theorem 6.29. *Any function computable by a Turing machine whose computation time is bounded by a primitive recursive function of the input length is primitive recursive.*

6.3 Simple Arithmetical Interpretation

Needless to say, we will not bother with representing algorithms by Turing machines. Neither will we be presenting exact estimates of complexity bounds. Such estimates may be useful for explaining how to do computation efficiently using limited resources. However, primitive recursive time bound is such a liberal restriction that one can usually establish it via a high-level observation of the algorithm. Roughly speaking, it suffices to make sure that all `while` cycles can be converted into `for` cycles executed a primitive recursive number of times.

Lemma 6.30. *Let* CS *be a primitive recursive and almost schematic constant specification. Let* $\mathcal{X} \subseteq \mathsf{Tm} \times \mathcal{L}_\mathsf{J}$ *be finite. Set* $\mathcal{B} := \mathsf{CS} \cup \mathcal{X}$. *Then the following binary relation on natural numbers is primitive recursive:*

$$\mathsf{evfun}_\mathcal{B} := \{(\ulcorner t \urcorner, \ulcorner G \urcorner) \mid (t, G) \in \mathcal{E}(\mathcal{B})\} \ .$$

Proof. We need to show that the characteristic function of $\mathsf{evfun}_\mathcal{B}$ is primitive recursive, i.e., we need an algorithm that, given $\ulcorner t \urcorner$ and $\ulcorner G \urcorner$, determines in primitive recursive time whether $(t, G) \in \mathcal{E}(\mathcal{B})$. According to (4.7), this is equivalent to determining whether G is an instance of some scheme $U \in \|t\|^{\mathsf{rk}(t)}$. The decision algorithm operating on terms and formulas directly in a suitable symbolic alphabet can be found in the proof of Lemma 4.30. The only difference now is that we need to operate on Gödel numbers of terms and formulas in the arithmetic setting. Given that all syntactic operations are primitive recursive by assumption, such an arithmetization of the algorithm does not present any difficulties. It does, however, require a more detailed description because objects, like sets of terms, that were considered basic earlier need to be encoded numerically. These details are quite mundane and are more useful as an exercise than as several pages of boring technical text to be read, so we omit them. The main observation that needs to be added to the proof of Lemma 4.30 is that cycles in that algorithm can generally be bounded by $\mathsf{rk}(t)$ iterations and that the corresponding function on Gödel numbers **trank** from (6.14) is primitive recursive. It should also be remarked that it is here where we need Gödel numbering of schemes of formulas of $\mathcal{L}_\mathsf{J}$. □

Next we need to show that the satisfaction relation for finitary CS-models is primitive recursive. Fortunately, it is sufficient to do it on a model-by-model basis, rather than going into the details of coding a finitary model numerically.

Theorem 6.31. *Let* CS *be a primitive recursive and almost schematic constant specification. For each finitary* CS*-model* $\mathcal{M}$*, the following unary relation on natural numbers is primitive recursive:*

$$\mathsf{satrel}_{\mathcal{M}} := \{\ulcorner A \urcorner \mid \mathcal{M} \Vdash A\} \ .$$

Proof. This theorem is obtained by arithmeticizing the proof of Theorem 4.31, except that we use Lemma 6.30 instead of Lemma 4.30. □

Last but not least, we need a function that, given the Gödel number of a term s, computes the set of Gödel numbers of all formulas F such that $\mathcal{M}_{\mathsf{fin}} \Vdash s{:}F$. More precisely, we need the following theorem.

Theorem 6.32. *Let* CS *be a finite constant specification and* $\mathcal{M} = (\mathsf{val}, \mathcal{B})$ *be a finitary* CS*-model. There is a primitive recursive function* $\mathsf{i}_{\mathcal{M}}(k)$ *such that,*

- *for each justification term s,*

$$\mathsf{i}_{\mathcal{M}}(\ulcorner s \urcorner) = \langle \ulcorner F_0 \urcorner, \ldots, \ulcorner F_{n-1} \urcorner \rangle$$

where $\{F_0, \ldots, F_{n-1}\} = \{F \mid \mathcal{M} \Vdash s{:}F\}$*, and*

- *for each k that is not a code of a justification term,* $\mathsf{i}_{\mathcal{M}}(k) = \langle \rangle$.

Proof. Once again, the proof relies essentially on the arithmeticized construction of the sets $\|t\|^i$ from Definition 4.25. In fact, the algorithm from the proof of Lemma 6.30 can be partially reused (a finite constant specification is both primitive recursive and almost schematic and $\mathcal{B}\setminus\mathsf{CS}$ must be finite by the definition of finitary models). Namely, the sequence

$$\langle \ulcorner G_0 \urcorner, \ldots, \ulcorner G_{m-1} \urcorner \rangle$$

where $\{G_0, \ldots, G_{m-1}\} = \{G \mid (s, G) \in \mathcal{E}(\mathcal{B})\} = \{G \mid G \in \|s\|^{\mathsf{rk}(s)}\}$ is constructed in the same way. Note that here all G_i's are formulas rather than schemas as in the proof of Lemma 6.30 because CS is finite and, hence, either not schematic or empty. It remains to remove from this sequence the Gödel numbers of those formulas G_i that are refuted in the model $\mathcal{M}$. For this we can use Theorem 6.31, which is applicable for largely the same reason as Lemma 6.30. □

Now all ingredients for this and the next section are ready and we come to the definition of a simple arithmetical interpretation.

6.3 Simple Arithmetical Interpretation

Definition 6.33 (Simple arithmetical interpretation). A *simple arithmetical interpretation* is a pair $(*, \mathsf{Prf})$ such that

1. $*$ maps atomic propositions of $\mathcal{L}_\mathsf{J}$ to sentences of $\mathcal{L}_\mathsf{PA}$;

2. $*$ maps evidence terms of $\mathcal{L}_\mathsf{J}$ to numerals $\underline{m}$ of $\mathcal{L}_\mathsf{PA}$;

3. Prf is a proof predicate for PA; and

4. for all evidence terms s and t we have:

$$\mathbb{N} \vDash \mathsf{Prf}(s^*, \ulcorner \phi \to \psi \urcorner) \wedge \mathsf{Prf}(t^*, \ulcorner \phi \urcorner) \to \mathsf{Prf}((s \cdot t)^*, \ulcorner \psi \urcorner) \quad (6.20)$$
$$\mathbb{N} \vDash \mathsf{Prf}(s^*, \ulcorner \phi \urcorner) \vee \mathsf{Prf}(t^*, \ulcorner \phi \urcorner) \to \mathsf{Prf}((s+t)^*, \ulcorner \phi \urcorner) \quad (6.21)$$
$$\mathbb{N} \vDash \mathsf{Prf}(s^*, \ulcorner \phi \urcorner) \to \mathsf{Prf}((!s)^*, \ulcorner \mathsf{Prf}(s^*, \ulcorner \phi \urcorner) \urcorner) \ . \quad (6.22)$$

We extend the mapping $*$ to all formulas of $\mathcal{L}_\mathsf{J}$ by setting

$$(t{:}F)^* := \mathsf{Prf}(t^*, \ulcorner F^* \urcorner) \qquad \bot^* := \bot \qquad (F \to G)^* := F^* \to G^* \ .$$

If there is no need to explicitly mention the proof predicate Prf, we denote the simple arithmetical interpretation $(*, \mathsf{Prf})$ by $*$.

A simple arithmetical interpretation $*$ is called *simple arithmetical CS-interpretation* if for each $(c, A) \in \mathsf{CS}$ we have $\mathbb{N} \vDash (c{:}A)^*$. An $\mathcal{L}_\mathsf{J}$-formula F is *weakly arithmetically CS-valid* if $\mathsf{PA} \vdash F^*$ for all simple arithmetical CS-interpretations $*$.

Remark 6.34. Note that we have started omitting underlining for certain numerals in the formulas above. This is not just to avoid double lines. The other reason is that both arguments of Prf in expressions $\mathsf{Prf}(t^*, \ulcorner \phi \urcorner)$ are numerals. However, formally speaking, t^* already represents a numeral by definition, whereas $\ulcorner \phi \urcorner$ represents a natural number and, hence, formally, requires an underline. However, there is a danger that underlining only one of the two arguments would create a false impression that only one of them is a numeral. Hence, we start implementing the simplified notation to avoid this potential confusion.

Theorem 6.35 (Simple Arithmetical Soundness). *Let CS be a constant specification and F be an $\mathcal{L}_\mathsf{J}$-formula. Then we have*

$$\mathsf{LP}_\mathsf{CS} \vdash F \quad \text{implies} \quad F \text{ is weakly arithmetically } \mathsf{CS}\text{-valid} \ .$$

Proof. Suppose $*$ is a simple arithmetical CS-interpretation. We proceed by induction on the $\mathsf{LP_{CS}}$-derivation of F. We use the fact that F^* is always a sentence.

If F is a classical tautology, then so is F^* and we trivially have $\mathsf{PA} \vdash F^*$.

If F is an instance of **j**, **j+**, or **j4**, then $\mathbb{N} \models F^*$ follows from (6.20)–(6.22). Since F is a Boolean combination of $s{:}G$-type formulas, F^* is a Boolean combination of substitution instances of provably Δ_1 formulas and is itself provably Δ_1 by Lemma 6.21. It follows by Theorem 6.23 that $\mathsf{PA} \vdash F^*$.

If F is the conclusion of an instance of axiom necessitation, then F has the form $c{:}A$ where $(c, A) \in \mathsf{CS}$. We have $\mathbb{N} \models (c{:}A)^*$ because $*$ is a CS-interpretation. Since $(c{:}A)^*$ is a substitution instance of a provably Δ_1 formula, by Lemma 6.21 it is provably Δ_1 itself, and

$$\mathsf{PA} \vdash (c{:}A)^*$$

by Theorem 6.23.

The only axiom instances F where we cannot be sure that F^* is provably Δ_1 are instances of **jt**. If $F = t{:}A \to A$, then

$$F^* = \mathsf{Prf}(t^*, \ulcorner A^* \urcorner) \to A^* \ .$$

To show that $\mathsf{PA} \vdash F^*$, we distinguish two cases, depending on whether the sentence $\mathsf{Prf}(t^*, \ulcorner A^* \urcorner)$ is true or false:

- $\mathbb{N} \models \mathsf{Prf}(t^*, \ulcorner A^* \urcorner)$. By (6.17), we find $\mathsf{PA} \vdash A^*$ and, thus,

$$\mathsf{PA} \vdash \mathsf{Prf}(t^*, \ulcorner A^* \urcorner) \to A^* \ .$$

- $\mathbb{N} \not\models \mathsf{Prf}(t^*, \ulcorner A^* \urcorner)$. In this case, $\mathsf{Prf}(t^*, \ulcorner A^* \urcorner)$ is a false provably Δ_1 sentence, meaning that $\mathsf{PA} \vdash \neg\mathsf{Prf}(t^*, \ulcorner A^* \urcorner)$ by Theorem 6.23. Therefore, again

$$\mathsf{PA} \vdash \mathsf{Prf}(t^*, \ulcorner A^* \urcorner) \to A^* \ .$$

Finally, if F is the conclusion of an instance of modus ponens, the claim follows by the induction hypothesis and the fact that $*$ distributes through implication. □

In the remainder of this section we show completeness of $\mathsf{LP_{CS}}$ with respect to any simple arithmetical CS-interpretation where CS is a primitive recursive and almost schematic constant specification. In order to obtain this result, we will establish the following property:

6.3 Simple Arithmetical Interpretation

Lemma 6.36. *For each finitary* CS*-model* $\mathcal{M}_{\mathsf{fin}}$, *there exists a simple arithmetical* CS*-interpretation* $*$ *such that for all* $\mathcal{L}_\mathsf{J}$*-formulas* G

$$\mathcal{M}_{\mathsf{fin}} \Vdash G \quad \text{implies} \quad \mathbb{N} \vDash G^* \ . \tag{6.23}$$

Simple arithmetical completeness then follows easily.

Theorem 6.37 (Simple arithmetical completeness). *Let* CS *be a primitive recursive and almost schematic constant specification. For any formula* F *of* $\mathcal{L}_\mathsf{J}$ *we have*

$$F \text{ is weakly arithmetically } \mathsf{CS}\text{-valid} \quad \text{implies} \quad \mathsf{LP}_{\mathsf{CS}} \vdash F \ .$$

Proof. Assume that $\mathsf{LP}_{\mathsf{CS}} \nvdash F$. By Theorem 4.35, there exists a finitary CS-model $\mathcal{M}_{\mathsf{fin}}$ with $\mathcal{M}_{\mathsf{fin}} \nVdash F$. Thus, $\mathcal{M}_{\mathsf{fin}} \Vdash \neg F$. By Lemma 6.36, there is a simple arithmetical CS-interpretation $*$ such that $\mathbb{N} \vDash (\neg F)^*$, i.e., $\mathbb{N} \vDash \neg(F^*)$. Therefore, $\mathbb{N} \nvDash F^*$, which implies $\mathsf{PA} \nvdash F^*$ by soundness (6.3) of PA. Hence, F is not weakly arithmetically CS-valid. $\square$

The complicated part is to establish (6.23). For the rest of this section, we assume that we are given

1. a constant specification CS that is primitive recursive and almost schematic,

2. a finitary CS-model $\mathcal{M}_{\mathsf{fin}}$.

Further, we recall that the Gödel numbering of the union of $\mathcal{L}_{\mathsf{PA}}$ and $\mathcal{L}_\mathsf{J}$ (with schemes) is injective, i.e.,

$$\ulcorner E_1 \urcorner = \ulcorner E_2 \urcorner \quad \text{if and only if} \quad E_1 = E_2$$

for any expressions E_1 and E_2.

We first have to decide which objects should serve as 'proofs' in our arithmetical interpretation. There will be two sources of 'proofs':

1. To begin with, all usual proofs will be 'proofs'. This guarantees that the direction from left to right in (6.17) is satisfied.

2. The second source of 'proofs' are the evidence terms of $\mathcal{L}_\mathsf{J}$. Every term t is a 'proof' for all formulas B for which $\mathcal{M}_{\mathsf{fin}} \Vdash t{:}B$.

To take care of the usual proofs, we make use of the usual primitive recursive proof predicate $\mathsf{Proof}(x,y)$ for Peano Arithmetic. Without loss of generality we assume that $\mathbb{N} \not\models \mathsf{Proof}(\ulcorner s \urcorner, k)$ for any evidence term s of $\mathcal{L}_\mathsf{J}$ and any natural number k.

In order to deal with the evidence terms, we denote by $\mathsf{Prf}(x,y)$ a formula with no free variables other than x and y that will be chosen later based on its desired properties that we are going to discuss now. For any such $\mathsf{Prf}(x,y)$, we can define an auxiliary translation † from $\mathcal{L}_\mathsf{J}$-formulas to $\mathcal{L}_\mathsf{PA}$-sentences as follows:

$$p^\dagger := \begin{cases} \ulcorner p \urcorner = \ulcorner p \urcorner & \text{if } \mathcal{M}_\mathsf{fin} \Vdash p \text{ ,} \\ \neg (\ulcorner p \urcorner = \ulcorner p \urcorner) & \text{otherwise} \end{cases} \quad \text{for any atomic proposition } p;$$

$$(t{:}F)^\dagger := \mathsf{Prf}\left(\ulcorner t \urcorner, \ulcorner F \urcorner\right) \text{ ;}$$

$$\bot^\dagger := \bot \text{ ;}$$

$$(F \to G)^\dagger := F^\dagger \to G^\dagger$$

(we give a complete notation in the definition). Obviously, atomic propositions that hold in $\mathcal{M}_\mathsf{fin}$ are translated to true provably Δ_1 sentences and atomic propositions that do not hold in $\mathcal{M}_\mathsf{fin}$ are translated to false provably Δ_1 sentences. The former are provable and the latter are refutable in PA.

Remark 6.38. We need the translation † to be injective. Therefore, simply putting

$$p^\dagger := \begin{cases} 0 = 0 & \text{if } \mathcal{M}_\mathsf{fin} \Vdash p \text{ ,} \\ 0 = 1 & \text{otherwise} \end{cases}$$

would not be sufficient.

Lemma 6.39. *If the formula* $\mathsf{Prf}(x,y)$ *contains some relation symbol $\underline{R}$ other than $=$, i.e., some relation symbol not occurring in the †-translation outside of* Prf*, then this translation is injective, in other words,*

$$F^\dagger = G^\dagger \quad \text{implies} \quad F = G \text{ .} \tag{6.24}$$

Proof. We assume $F^\dagger = G^\dagger$ and show (6.24) by induction on the structure of the $\mathcal{L}_\mathsf{J}$-formula F.

(i) F is an atomic proposition. By the definition of †, G must also be an atomic proposition and, by the injectivity of the Gödel numbering, G must be the same atomic proposition as F.

6.3 Simple Arithmetical Interpretation

(ii) F is $\bot$. By the definition of $\dagger$, it is clear that $G = \bot$.

(iii) F is a formula $s{:}F_1$. Then G must be of the form $t{:}G_1$. Indeed, by the definition of $\dagger$, it is obvious that G can be neither $\bot$ nor an atomic proposition. Suppose towards a contradiction the only remaining possibility, i.e., that $G = G_1 \to G_2$. Since $(s{:}F_1)^\dagger = \mathsf{Prf}(\underline{k},\underline{n})$ for suitable k and n, the sentence $(s{:}F_1)^\dagger = G_1{}^\dagger \to G_2{}^\dagger$ would contain the symbol $\underline{R}$, meaning that $G_1{}^\dagger$ or $G_2{}^\dagger$ would contain a subformula of the form $\mathsf{Prf}(\underline{k_1},\underline{n_1})$. Let l be the number of occurrences of $\to$ in $\mathsf{Prf}(x,y)$. Substitutions of terms for variables do not affect Boolean connectives, hence both $(s{:}F_1)^\dagger$ and $\mathsf{Prf}(\underline{k_1},\underline{n_1})$ would contain exactly l occurrences of $\to$ each. But then $G_1{}^\dagger \to G_2{}^\dagger$ would contain at least $l+1$ occurrences of $\to$. This contradiction shows the impossibility of $(s{:}F_1)^\dagger = G_1{}^\dagger \to G_2{}^\dagger$. Therefore, $G = t{:}G_1$. By the induction hypothesis and injectivity of the Gödel numbering we conclude $s = t$ and $F_1 = G_1$.

(iv) F is $F_1 \to F_2$. By the same argument as in (iii), we have

$$G = G_1 \to G_2 \ .$$

By the induction hypothesis, $F_1 = G_1$ and $F_2 = G_2$. □

For any formula $\mathsf{Prf}(x,y)$ that yields an injective $\dagger$, it can be shown by using the standard techniques for Gödel numbering that binary functions $\mathsf{dag}(x,y)$ and $\mathsf{undag}(x,y)$ such that

$$\mathsf{dag}(\ulcorner B \urcorner, \ulcorner \mathsf{Prf}(x,y) \urcorner) = \ulcorner B^\dagger \urcorner , \tag{6.25}$$
$$\mathsf{undag}(\ulcorner B^\dagger \urcorner, \ulcorner \mathsf{Prf}(x,y) \urcorner) = \ulcorner B \urcorner \tag{6.26}$$

are primitive recursive and our language contains the corresponding function symbols $\underline{\mathsf{dag}}$ and $\underline{\mathsf{undag}}$ (as usual, it does not matter much how these functions are defined on inputs that are not Gödel numbers of such formulas, e.g., they can be assumed to be constant on all other inputs). Note that the functions dag and undag are supposed to take the Gödel number of the formula $\mathsf{Prf}(x,y)$ as a parameter. Hence, unlike the translation $\dagger$, these functions do not depend on $\mathsf{Prf}(x,y)$. This means, in particular, that the way undag is defined does not depend on whether $\dagger$ is injective. Property (6.26), however, is only guaranteed for injective $\dagger$'s. Note also that dag and undag do depend on the chosen model $\mathcal{M}_{\mathsf{fin}}$.

By Theorem 6.31, the satisfaction relation for $\mathcal{M}_{\mathsf{fin}}$ is primitive recursive. Therefore, so is the binary relation

$$\mathsf{Jus} := \{(\ulcorner s \urcorner, \ulcorner F \urcorner) \mid \mathcal{M}_{\mathsf{fin}} \Vdash s{:}F\} \ ,$$

and our the language of PA has the corresponding binary relation symbol $\underline{\mathsf{Jus}}$. By Lemma 6.21, $\underline{\mathsf{Jus}}(t_1, t_2)$ is a provably Δ_1 formula for arbitrary terms t_1 and t_2. Using Diagonalization Lemma 6.24, we now define the desired formula $\mathsf{Prf}(x, y)$ to satisfy

$$\mathsf{PA} \vdash \mathsf{Prf}(x,y) \leftrightarrow \mathsf{Proof}(x,y) \vee \underline{\mathsf{Jus}}\Big(x, \mathsf{undag}(y, \ulcorner \mathsf{Prf}(x,y) \urcorner)\Big) \ . \quad (6.27)$$

Moreover, since $\mathsf{Proof}(x,y) \vee \underline{\mathsf{Jus}}(x, \mathsf{undag}(y,z))$ is clearly provably Δ_1, so is our $\mathsf{Prf}(x,y)$ by Lemma 6.24. It follows by Lemma 6.21 that $F^\dagger$ is provably Δ_1 for each F.

Thus, by soundness (6.3) of PA, for the universal closure of (6.27),

$$\mathbb{N} \models \forall x \forall y \Big(\mathsf{Prf}(x,y) \leftrightarrow \mathsf{Proof}(x,y) \vee \underline{\mathsf{Jus}}\Big(x, \mathsf{undag}(y, \ulcorner \mathsf{Prf}(x,y) \urcorner)\Big)\Big) \ . \quad (6.28)$$

Further, (6.24) holds by Lemma 6.24 because the formula $\mathsf{Prf}(x,y)$ contains the relation symbol $\underline{\mathsf{Jus}}$. It follows that undag really performs the inverse translation, i.e., informally we have

$$\mathsf{Prf}(x,y) \quad \Longleftrightarrow \quad \begin{array}{l} \mathsf{Proof}(x,y) \\ \text{or} \\ \text{there is a term } s \text{ and a formula } F \text{ such that} \\ x = \ulcorner s \urcorner, \text{ and } y = \ulcorner F^\dagger \urcorner, \text{ and } \mathcal{M}_{\mathsf{fin}} \Vdash s{:}F \ . \end{array}$$

The key property of the translation † that is based on the chosen Prf is that $\mathcal{L}_\mathsf{J}$-formulas that hold (do not hold) in $\mathcal{M}_{\mathsf{fin}}$ are translated to true (false) $\mathcal{L}_{\mathsf{PA}}$-sentences, as stated formally in the following lemma:

Lemma 6.40. *For each formula F of $\mathcal{L}_\mathsf{J}$,*

(i) $\mathcal{M}_{\mathsf{fin}} \Vdash F$ *implies* $\mathbb{N} \models F^\dagger$;

(ii) $\mathcal{M}_{\mathsf{fin}} \not\Vdash F$ *implies* $\mathbb{N} \not\models F^\dagger$.

Proof. By simultaneous induction on the structure of F. We distinguish the following cases:

6.3 Simple Arithmetical Interpretation 175

(i) Let F be an atomic proposition. If $\mathcal{M}_{\text{fin}} \Vdash F$, then $F^\dagger$ is

$$\ulcorner F \urcorner = \ulcorner F \urcorner ,$$

which clearly is true. If $\mathcal{M}_{\text{fin}} \nVdash F$, then $F^\dagger$ is $\neg(\ulcorner F \urcorner = \ulcorner F \urcorner)$, which clearly is false.

(ii) If $F = \bot$, then trivially we have $\mathcal{M}_{\text{fin}} \nVdash \bot$ and $\mathbb{N} \nvDash \bot$.

(iii) The case of $F = G \to H$ is immediate by the induction hypothesis.

(iv) Let $F = s{:}G$. If $\mathcal{M}_{\text{fin}} \Vdash s{:}G$, then $\mathbb{N} \vDash \text{\underline{Jus}}(\ulcorner s \urcorner, \ulcorner G \urcorner)$. Since by (6.26) this $\ulcorner G \urcorner = \text{undag}\left(\ulcorner G^\dagger \urcorner, \ulcorner \text{Prf}(x,y) \urcorner\right)$, the second disjunct of the fixed point definition in (6.28) is made true for the instantiation of x with $\ulcorner s \urcorner$ and y with $\ulcorner G^\dagger \urcorner$. Therefore, $\mathbb{N} \vDash \text{Prf}\left(\ulcorner s \urcorner, \ulcorner G^\dagger \urcorner\right)$, i.e., $\mathbb{N} \vDash (s{:}G)^\dagger$.

If $\mathcal{M}_{\text{fin}} \nVdash s{:}G$, then $\mathbb{N} \nvDash \text{\underline{Jus}}(\ulcorner s \urcorner, \ulcorner G \urcorner)$, which makes the same instantiation of the second disjunct of (6.28) false. Moreover, the first disjunct for the same instantiation is also false, $\mathbb{N} \nvDash \text{Proof}\left(\ulcorner s \urcorner, \ulcorner G^\dagger \urcorner\right)$, because by our assumptions $\mathbb{N} \nvDash \text{Proof}(\ulcorner s \urcorner, \underline{k})$ for any k. Thus, in this case, we have $\mathbb{N} \nvDash \text{Prf}\left(\ulcorner s \urcorner, \ulcorner G^\dagger \urcorner\right)$, i.e., $\mathbb{N} \nvDash (s{:}G)^\dagger$. $\square$

We have already established that $\text{Prf}(x,y)$ is a provably Δ_1 formula. To show that $\text{Prf}(x,y)$ is a proof predicate it remains to prove the following lemma:

Lemma 6.41. *For every $\mathcal{L}_{\text{PA}}$-sentence ϕ,*

$$\text{PA} \vdash \phi \quad \Longleftrightarrow \quad \mathbb{N} \vDash \text{Prf}(\underline{n}, \ulcorner \phi \urcorner) \text{ for some natural number } n.$$

Proof. From left to right. Suppose $\text{PA} \vdash \phi$. Then there is a natural number n such that $\mathbb{N} \vDash \text{Proof}(\underline{n}, \ulcorner \phi \urcorner)$ because $\text{Proof}(x,y)$ is a proof predicate. By (6.28), we conclude $\mathbb{N} \vDash \text{Prf}(\underline{n}, \ulcorner \phi \urcorner)$.

From right to left. Suppose that $\mathbb{N} \vDash \text{Prf}(\underline{n}, \ulcorner \phi \urcorner)$. Then, by (6.28), either $\mathbb{N} \vDash \text{Proof}(\underline{n}, \ulcorner \phi \urcorner)$, in which case $\text{PA} \vdash \phi$ follows immediately, or $n = \ulcorner s \urcorner$ for some evidence term s and some $\mathcal{L}_\text{J}$-formula F such that

$$\ulcorner F \urcorner = \text{undag}\left(\ulcorner \phi \urcorner, \ulcorner \text{Prf}(x,y) \urcorner\right) \quad \text{and} \quad \mathcal{M}_{\text{fin}} \Vdash s{:}F .$$

Therefore, $\phi = F^\dagger$ and $\mathcal{M}_{\text{fin}} \Vdash F$. By the previous lemma, $\mathbb{N} \vDash F^\dagger$. Since $F^\dagger$ is a provably Δ_1 sentence, we find $\text{PA} \vdash F^\dagger$, i.e., $\text{PA} \vdash \phi$. $\square$

Now we obtain a simple arithmetical CS-interpretation as follows.

Lemma 6.42. *Let $*$ be a mapping such that $s^* := \ulcorner\underline{s}\urcorner$ for each evidence term s and $p^* := p^\dagger$ for each atomic proposition p. Then the pair $(*, \mathsf{Prf})$ is a simple arithmetical CS-interpretation. Moreover,*

$$F^* = F^\dagger \tag{6.29}$$

for any $\mathcal{L}_\mathsf{J}$-formula F.

Proof. We start with showing (6.29) by induction on the structure of F. We distinguish the following cases.

(i) If F is an atomic proposition, then $F^* = F^\dagger$ by definition.

(ii) If $F = t{:}G$, then $t^* = \ulcorner\underline{t}\urcorner$. By the induction hypothesis, $G^* = G^\dagger$. Thus,
$$(t{:}G)^* = \mathsf{Prf}(t^*, \ulcorner G^*\urcorner) = \mathsf{Prf}(\ulcorner\underline{t}\urcorner, \ulcorner G^\dagger\urcorner) = (t{:}G)^\dagger \;.$$

(iii) If $F = \bot$, then $\bot^* = \bot^\dagger$ by definition.

(iv) If $F = G \to H$, then $G^* = G^\dagger$ and $H^* = H^\dagger$ by the induction hypothesis. Thus, $(G \to H)^* = G^* \to H^* = G^\dagger \to H^\dagger = (G \to H)^\dagger$.

This finishes the proof of (6.29).

Now we show that $(*, \mathsf{Prf})$ is a simple arithmetical CS-interpretation. The mapping $*$ maps atomic propositions of $\mathcal{L}_\mathsf{J}$ to sentences of $\mathcal{L}_\mathsf{PA}$ and evidence terms to numerals. Further, Prf is a proof predicate by the previous lemma. It remains to establish (6.20)–(6.22) from Definition 6.33. We only present a proof of (6.20). The other proofs are similar.

Assume that $\mathbb{N} \models \mathsf{Prf}(s^*, \ulcorner\phi \to \psi\urcorner)$ and $\mathbb{N} \models \mathsf{Prf}(t^*, \ulcorner\phi\urcorner)$, i.e., in other words, $\mathbb{N} \models \mathsf{Prf}(\ulcorner\underline{s}\urcorner, \ulcorner\phi \to \psi\urcorner)$ and $\mathbb{N} \models \mathsf{Prf}(\ulcorner\underline{t}\urcorner, \ulcorner\phi\urcorner)$. By our assumptions, $\mathbb{N} \not\models \mathsf{Proof}(\ulcorner r\urcorner, \underline{k})$ for any evidence term r and any natural number k. Therefore, by (6.28), we find $\phi = F^\dagger$ and $\psi = G^\dagger$ for some $\mathcal{L}_\mathsf{J}$-formulas F and G such that

$$\mathcal{M}_\mathsf{fin} \Vdash s{:}(F \to G) \quad \text{and} \quad \mathcal{M}_\mathsf{fin} \Vdash t{:}F \;.$$

Hence, $\mathcal{M}_\mathsf{fin} \Vdash (s \cdot t){:}G$. Therefore, by (6.28), we obtain

$$\mathbb{N} \models \mathsf{Prf}(\ulcorner\underline{s \cdot t}\urcorner, \ulcorner G^\dagger\urcorner) \;,$$

which is $\mathbb{N} \models \mathsf{Prf}\big((s \cdot t)^*, \ulcorner\psi\urcorner\big)$.

6.4 Arithmetical Interpretation

It remains to show that the constant specification is respected. Let $(c, A) \in \mathsf{CS}$. Then $\mathcal{M}_{\mathsf{fin}} \Vdash c{:}A$. Thus, by Lemma 6.40, we have

$$\mathbb{N} \vDash (c{:}A)^\dagger \ .$$

Hence, by (6.29), we have $\mathbb{N} \vDash (c{:}A)^*$. □

Now (6.23) follows easily. First observe that by Lemma 6.42 the pair $(*, \mathsf{Prf})$ is a simple arithmetical CS-interpretation. Suppose $\mathcal{M}_{\mathsf{fin}} \Vdash G$. By Lemma 6.40 we find $\mathbb{N} \vDash G^\dagger$, which is $\mathbb{N} \vDash G^*$ by (6.29). This closes the proof of simple arithmetical completeness.

6.4 Arithmetical Interpretation

Weak arithmetical interpretations do not provide a semantics for the term operations of the Logic of Proofs. This is remedied in the arithmetical interpretations studied in this section.

For the purpose of this section, it is important that proof predicates $\mathsf{Proof}(x, y)$ are multi-conclusion.[14] That is for a given natural number n, there may be several different formulas ϕ_i with $\mathbb{N} \vDash \mathsf{Proof}(\underline{n}, \ulcorner \phi_i \urcorner)$.

We can obtain such a proof predicate as follows. Let $\mathsf{isProof}(k)$ be a unary primitive recursive relation such that $\mathsf{isProof}(k)$ if and only if

1. k encodes a finite sequence $\langle (k)_0, (k)_1, \ldots, (k)_{\mathsf{len}(k)-1} \rangle$,

2. each $(k)_i$ is the Gödel number of some $\mathcal{L}_{\mathsf{PA}}$-formula, and

3. if $(k)_i = \ulcorner A \urcorner$, then either A is an axiom of PA or A can be inferred with a single rule application in PA from formula(s) A_j such that for all j we have $\ulcorner A_j \urcorner = (k)_{l_j}$ for some $l_j < i$.

We now define a binary primitive recursive relation $\mathsf{Proof}(k, l)$ by

$$\mathsf{Proof}(k, l) := \mathsf{isProof}(k) \wedge \left(\exists i < \mathsf{len}(k)\right) l = (k)_i \ .$$

Thus, $\mathsf{Proof}(k, l)$ means

"k is the code of a proof containing a formula with code l."

---

[14] Predicates Prf were multi-conclusion already in the preceding section, mostly due to the constants generally justifying more than one axiom. However, there it was not important whether the standard predicate Proof is too.

Obviously, $\underline{\mathsf{Proof}}(x,y)$ is a proof predicate. Note that this proof predicate is multi-conclusion but finitely so: for each k there only finitely many formulas ϕ_i such that $\mathsf{Proof}(k, \ulcorner \phi_i \urcorner)$, or, equivalently, such that $\mathbb{N} \vDash \underline{\mathsf{Proof}}(\underline{k}, \ulcorner \phi_i \urcorner)$.

Remark 6.43. In the preceding section on simple arithmetical interpretations, it did not matter which standard proof predicate $\mathsf{Proof}(x,y)$ was used, which is why we did not specify it. In this section, we need to know the exact structure of the standard proof predicate and will be using $\underline{\mathsf{Proof}}(x,y)$ just constructed above. Since all the results of the preceding section apply to it too, we retcon the preceding section by claiming that the multi-conclusion proof predicate from this section has been always used whenever we talked about Proof. Thus, all the results about Proof from the preceding section, including (6.27) for each finitary CS-model $\mathcal{M}_{\mathsf{fin}}$ for a primitive recursive and almost schematic CS, hold.

Definition 6.44 (Normal proof predicate). A *normal proof predicate* is a quadruple $(\mathsf{Prf}(x,y), \tilde{\otimes}, \tilde{\oplus}, \tilde{\mathsf{e}})$ such that

1. $\mathsf{Prf}(x,y)$ is a proof predicate and

2. $\tilde{\otimes}, \tilde{\oplus}$, and $\tilde{\mathsf{e}}$ are recursive functions such that for all natural numbers m and n and for all $\mathcal{L}_{\mathsf{PA}}$-formulas ϕ and ψ,

$$\mathbb{N} \vDash \mathsf{Prf}(\underline{m}, \ulcorner \phi \to \psi \urcorner) \land \mathsf{Prf}(\underline{n}, \ulcorner \phi \urcorner) \to \mathsf{Prf}(\underline{m \tilde{\otimes} n}, \ulcorner \psi \urcorner) \;, \quad (6.30)$$
$$\mathbb{N} \vDash \mathsf{Prf}(\underline{m}, \ulcorner \phi \urcorner) \lor \mathsf{Prf}(\underline{n}, \ulcorner \phi \urcorner) \to \mathsf{Prf}(\underline{m \tilde{\oplus} n}, \ulcorner \phi \urcorner) \;, \quad (6.31)$$
$$\mathbb{N} \vDash \mathsf{Prf}(\underline{n}, \ulcorner \phi \urcorner) \to \mathsf{Prf}\left(\underline{\tilde{\mathsf{e}}(n)}, \ulcorner \mathsf{Prf}(\underline{n}, \ulcorner \phi \urcorner) \urcorner \right) \;. \quad (6.32)$$

Next we define three total recursive[15] functions $\otimes$, $\oplus$, and e such that the quadruple $(\underline{\mathsf{Proof}}(x,y), \otimes, \oplus, \mathsf{e})$ is a normal proof predicate.

Definition 6.45 (Functions $\oplus$, $\otimes$, and e for Proof). The function

$$\oplus := \mathsf{concat}$$

simply concatenates two given sequences.

The function $\otimes$ takes the result of using $\oplus$ and appends to it all $\ulcorner \phi \urcorner$ for which there is a formula ψ such that both $\ulcorner \psi \urcorner$ and $\ulcorner \psi \to \phi \urcorner$ are present in this result.

---

[15] Such functions could even be defined in a primitive recursive way, but imposing a stronger restriction would not make sense as it would have to be relaxed later anyway.

6.4 Arithmetical Interpretation

It remains to provide a function e. Since $\underline{\mathsf{Proof}}(x,y)$ is a proof predicate and, in particular, a provably Δ_1 formula, by Theorem 6.23 and Lemma 6.21 we have for arbitrary natural numbers m and n that

$$\mathbb{N} \vDash \underline{\mathsf{Proof}}(\underline{n}, \underline{m}) \implies \mathbb{N} \vDash \underline{\mathsf{Proof}}\left(\underline{k}, \ulcorner \underline{\mathsf{Proof}}(\underline{n}, \underline{m})\urcorner\right) \text{ for some } k.$$

Equivalently,

$$\mathsf{Proof}(n, m) \implies \mathsf{Proof}\left(k, \ulcorner \underline{\mathsf{Proof}}(\underline{n}, \underline{m})\urcorner\right) \text{ for some } k. \qquad (6.33)$$

Using the primitive recursive functions (6.11) and (6.12), it is easy to show that $\ulcorner \underline{\mathsf{Proof}}(\underline{n}, \underline{m}) \urcorner$ is a binary primitive recursive function of n and m. Thus,

$$R(k, n, m) := \mathsf{Proof}(n, m) \to \mathsf{Proof}\left(k, \ulcorner \underline{\mathsf{Proof}}(\underline{n}, \underline{m}) \urcorner\right)$$

is a ternary primitive recursive relation. We define

$$f(n, m) := [\mu.\chi_R](n, m) \ ,$$

which is a binary total recursive function due to (6.33). Finally, we define

$$\mathsf{e}(n) := f\bigl(n, (n)_0\bigr) \oplus \cdots \oplus f\bigl(n, (n)_{\mathsf{len}(n)-1}\bigr) \ ,$$

which is a total recursive function because it uses a number of iterations of $\oplus$ bounded by a primitive recursive function.

We immediately get the following result.

Lemma 6.46. $(\mathsf{Proof}(x, y), \otimes, \oplus, \mathsf{e})$ *is a normal proof predicate.*

Proof. The only thing we need to establish is Properties (6.30)–(6.32) for $\oplus$, $\otimes$, and e. While the first two of these properties follow from the definitions rather straightforwardly, the construction of e is less trivial, so let us provide the argument for (6.32). Assume that $\mathbb{N} \vDash \underline{\mathsf{Proof}}(\underline{n}, \ulcorner\phi\urcorner)$, in other words, $\mathsf{Proof}(n, \ulcorner\phi\urcorner)$. This means, in particular, that

$$\ulcorner\phi\urcorner = (n)_i \text{ for some number } i < \mathsf{len}(n).$$

By (6.33), it follows that $f\bigl(n, (n)_i\bigr)$ is a number k such that

$$\mathsf{Proof}\left(k, \ulcorner \underline{\mathsf{Proof}}(\underline{n}, \ulcorner\phi\urcorner) \urcorner\right) \ .$$

This means that $f\bigl(n, (n)_i\bigr)$ is a code of a sequence that contains the code $\ulcorner \underline{\mathsf{Proof}}(\underline{n}, \ulcorner\phi\urcorner) \urcorner$. Since the sequence $\mathsf{e}(n)$ contains $f\bigl(n, (n)_i\bigr)$ as a subsequence, the former also contains this code and, hence,

$$\mathsf{Proof}\left(\mathsf{e}(n), \ulcorner \underline{\mathsf{Proof}}(\underline{n}, \ulcorner\phi\urcorner) \urcorner\right) \ . \qquad \square$$

Definition 6.47 (Arithmetical interpretation). An *arithmetical interpretation* is a quintuple
$$(*, \mathsf{Prf}, \tilde{\otimes}, \tilde{\oplus}, \tilde{\mathsf{e}})$$
such that

1. $*$ maps atomic propositions of $\mathcal{L}_\mathsf{J}$ to sentences of $\mathcal{L}_\mathsf{PA}$;

2. $*$ maps atomic evidence terms of $\mathcal{L}_\mathsf{J}$ to natural numbers;

3. $(\mathsf{Prf}, \tilde{\otimes}, \tilde{\oplus}, \tilde{\mathsf{e}})$ is a normal proof predicate.

We extend the mapping $*$ to all evidence terms of $\mathcal{L}_\mathsf{J}$ by setting
$$(s \cdot t)^* := s^* \tilde{\otimes} t^* \qquad (s+t)^* := s^* \tilde{\oplus} t^* \qquad (!t)^* := \tilde{\mathsf{e}}(t^*) \ . \qquad (6.34)$$

Then we extend the mapping $*$ to all formulas of $\mathcal{L}_\mathsf{J}$ by setting
$$(t{:}F)^* := \mathsf{Prf}(\underline{t^*}, \ulcorner F^* \urcorner) \qquad \bot^* := \bot \qquad (F \to G)^* := F^* \to G^* \ .$$

As before, we sometimes use just $*$ to denote the arithmetical interpretation $(*, \mathsf{Prf}, \tilde{\otimes}, \tilde{\oplus}, \tilde{\mathsf{e}})$ if there is no need to explicitly mention the normal proof predicate $(\mathsf{Prf}, \tilde{\otimes}, \tilde{\oplus}, \tilde{\mathsf{e}})$.

Suppose we are given a constant specification CS. An arithmetical interpretation $*$ is called an *arithmetical* CS-*interpretation* if for each $(c, A) \in \mathsf{CS}$ we have $\mathbb{N} \vDash (c{:}A)^*$. An $\mathcal{L}_\mathsf{J}$-formula F is *arithmetically* CS-*valid* if $\mathsf{PA} \vdash F^*$ for all arithmetical CS-interpretations $*$.

Remark 6.48. Let $(*, \mathsf{Prf}, \tilde{\otimes}, \tilde{\oplus}, \tilde{\mathsf{e}})$ be an arithmetical CS-interpretation. Define $t^{*_s} := \underline{t^*}$ and $p^{*_s} := p^*$. By (6.34) and (6.30)–(6.32) we see that the proof predicate $\mathsf{Prf}(x,y)$ and the mapping $*_s$ satisfy (6.20)–(6.22). Since clearly $F^{*_s} = F^*$ for each formula F, we can regard $*$ also as a simple arithmetical CS-interpretation.[16]

As a corollary of this remark we obtain soundness of LP_CS with respect to arithmetical interpretations.

---

[16]There is a bit of a mismatch in how terms are treated in arithmetical and simple arithmetical interpretations: here terms are translated into numbers that need to be turned into numerals when plugged into Prf, whereas there terms were interpreted as numerals from the start. This is, however, completely inconsequential. This technical deviation occurred because we wanted to simplify the notation in the case of simple interpretations.

6.4 Arithmetical Interpretation

Corollary 6.49 (Arithmetical Soundness). *Let* CS *be any constant specification and F be any $\mathcal{L}_\mathsf{J}$-formula. Then*

$$\mathsf{LP}_\mathsf{CS} \vdash F \quad \textit{implies} \quad F \textit{ is arithmetically } \mathsf{CS}\textit{-valid} \ .$$

The remainder of this section is devoted to establishing completeness of LP_CS with respect to arithmetical interpretations. However, we will only be able to show completeness in the case of finite constant specifications.[17] Hence for the rest of this section, we assume that CS is a given finite constant specification. Note that for a finite constant specification the results from the previous section still hold since a finite constant specification is, by definition, almost schematic and primitive recursive.

Again, we first show that for each finitary CS-model $\mathcal{M}_\mathsf{fin}$, there is an arithmetical CS-interpretation $(*, \mathsf{Prf}, \tilde{\otimes}, \tilde{\oplus}, \tilde{\mathsf{e}})$ such that for all $\mathcal{L}_\mathsf{J}$-formulas G

$$\mathcal{M}_\mathsf{fin} \Vdash G \quad \text{implies} \quad \mathbb{N} \vDash G^* \ . \tag{6.35}$$

So for the rest of this section, we assume that we are given a finitary CS-model $\mathcal{M}_\mathsf{fin}$. Then there is a simple arithmetical interpretation $(*_s, \mathsf{Prf})$ satisfying (6.23) and (6.27). We define $t^* := (t^{*_s})^\mathbb{N}$ for each atomic evidence term t and $p^* := p^{*_s}$ for each atomic proposition p. In order to show (6.35), it is sufficient to define partial recursive functions $\tilde{\otimes}$, $\tilde{\oplus}$, and $\tilde{\mathsf{e}}$ such that $(\mathsf{Prf}, \tilde{\otimes}, \tilde{\oplus}, \tilde{\mathsf{e}})$ is a normal proof predicate and $F^* = F^{*_s}$ for each formula $F \in \mathcal{L}_\mathsf{J}$.

To define these functions, we first need some auxiliary functions that model the construction of evidence terms on the level of Gödel numbers. Let $\overline{+}$, $\overline{\times}$, and $\overline{!}$ stand for primitive recursive functions such that for arbitrary evidence terms s and t

$$\ulcorner s \urcorner \mathbin{\overline{+}} \ulcorner t \urcorner = \ulcorner s + t \urcorner$$
$$\ulcorner s \urcorner \mathbin{\overline{\times}} \ulcorner t \urcorner = \ulcorner s \cdot t \urcorner$$
$$\overline{!}(\ulcorner s \urcorner) = \ulcorner !s \urcorner \ .$$

By Theorem 6.32, there exists a unary primitive recursive function $\mathsf{i}_{\mathcal{M}_\mathsf{fin}}(k)$ such that, for each evidence term s,

$$\mathsf{i}_{\mathcal{M}_\mathsf{fin}}(\ulcorner s \urcorner) = \langle \ulcorner F_0 \urcorner, \ldots, \ulcorner F_{n-1} \urcorner \rangle$$

---

[17] It is an open question whether this is a limitation of the proof or whether completeness can fail without the finiteness condition.

where $\{F_0, \ldots, F_{n-1}\} = \{F \mid \mathcal{M}_{\mathsf{fin}} \Vdash s{:}F\}$ and $\mathrm{i}_{\mathcal{M}_{\mathsf{fin}}}(k) = \langle\rangle$ if k is not a code of an evidence term.

Accordingly, the following binary relation

$$R_t(l,k) := \bigl(\forall j < \mathsf{len}(\mathrm{i}_{\mathcal{M}_{\mathsf{fin}}}(k))\bigr)\mathsf{Proof}\Bigl(l, \mathsf{dag}\bigl((\mathrm{i}_{\mathcal{M}_{\mathsf{fin}}}(k))_j, \ulcorner\mathsf{Prf}(x,y)\urcorner\bigr)\Bigr)$$

is primitive recursive. We define a unary recursive function

$$\mathsf{h}(k) := [\mu.\chi_{R_t}](k) \ .$$

That is for any evidence term s we have that $\mathsf{h}(\ulcorner s \urcorner)$ is the least l such that

$$\mathsf{Proof}(l, \ulcorner F^\dagger \urcorner) \quad \text{whenever } \mathcal{M}_{\mathsf{fin}} \Vdash s{:}F \ .$$

Note that such an l always exists. Indeed, suppose $\mathcal{M}_{\mathsf{fin}} \Vdash s{:}F$. Then $\mathcal{M}_{\mathsf{fin}} \Vdash F$. Thus by Lemma 6.40 we have $\mathbb{N} \models F^\dagger$ and, hence, $\mathsf{PA} \vdash F^\dagger$ (recall that $F^\dagger$ is always a provably Δ_1 sentence). Therefore, there exists an l_F such that $\mathsf{Proof}(l_F, \ulcorner F^\dagger \urcorner)$. By Theorem 6.32, there are only finitely many F_i's ($0 \le i < n$) such that $\mathcal{M}_{\mathsf{fin}} \Vdash s{:}F_i$. We can concatenate the corresponding sequences l_{F_i} by $l' := l_{F_0} \oplus \cdots \oplus l_{F_{n-1}}$, which yields

$$\mathsf{Proof}(l', \ulcorner F^\dagger \urcorner) \quad \text{whenever } \mathcal{M}_{\mathsf{fin}} \Vdash s{:}F \ .$$

To show that $\mathsf{h}(k)$ is a total recursive function, it remains to consider inputs k that are not codes of evidence terms. For them $\mathrm{i}_{\mathcal{M}_{\mathsf{fin}}}(k) = \langle\rangle$ is the empty sequence, whose length is 0. Thus, the bounded quantifier in the definition of $R(l,k)$ makes it vacuously true for all l, meaning that $\mathsf{h}(k) = 0$.

By (6.27), for arbitrary natural numbers n and m,

$$\mathsf{PA} \vdash \underline{\mathsf{Proof}}(\underline{n}, \underline{m}) \to \mathsf{Prf}(\underline{n}, \underline{m}) \ .$$

Similar to R_t, we can construct a binary primitive recursive relation

$$R_a(l, n) := \bigl(\forall j < \mathsf{len}(n)\bigr) \, \mathsf{Proof}\Bigl(l, \ulcorner \underline{\mathsf{Proof}}\bigl(\underline{n}, \underline{(n)_j}\bigr) \to \mathsf{Prf}\bigl(\underline{n}, \underline{(n)_j}\bigr) \urcorner\Bigr)$$

and define a unary recursive function $\mathsf{g}(n) := [\mu.\chi_{R_a}](n)$.[18] In other words, for any code $n = \langle \ulcorner\phi_0\urcorner, \ldots, \ulcorner\phi_{\mathsf{len}(n)-1}\urcorner\rangle$ of a sequence of codes of $\mathcal{L}_{\mathsf{PA}}$-formulas[19], we have that $\mathsf{g}(n)$ is the least l such that

$$\mathsf{Proof}\Bigl(l, \ulcorner \underline{\mathsf{Proof}}\bigl(\underline{n}, \ulcorner\phi_j\urcorner\bigr) \to \mathsf{Prf}\bigl(\underline{n}, \ulcorner\phi_j\urcorner\bigr) \urcorner\Bigr)$$

---

[18]It may be undefined on inputs n that are not (codes of) sequences if $\mathsf{len}(n) \ne 0$. Recall that we did not specify the behavior of len on (the codes of) non-sequences.

[19]In fact, it is sufficient that n be (the code of) a sequence of arbitrary numbers.

6.4 Arithmetical Interpretation

for all $0 \leq j < \mathsf{len}(n)$. Again, such l exists for each sequence (of formulas). We conclude that for each $n = \langle \ulcorner \phi_0 \urcorner, \ldots, \ulcorner \phi_{\mathsf{len}(n)-1} \urcorner \rangle$ the recursive function $\mathsf{g}(n)$ is defined and

$$\mathsf{Proof}\Big(\mathsf{g}(n), \ulcorner \underline{\mathsf{Proof}}\left(\underline{n}, \ulcorner \phi_j \urcorner\right) \to \mathsf{Prf}\left(\underline{n}, \ulcorner \phi_j \urcorner\right) \urcorner \Big) \tag{6.36}$$

for all $0 \leq j < \mathsf{len}(n)$.

We define a binary function $\tilde{\oplus}$ by

$$k \mathbin{\tilde{\oplus}} l := \begin{cases} k \oplus l & \text{if isProof}(k) \text{ and isProof}(l), \\ k \oplus \mathsf{h}(l) & \text{if isProof}(k) \text{ and } \overline{\mathsf{isProof}}(l), \\ \mathsf{h}(k) \oplus l & \text{if } \overline{\mathsf{isProof}}(k) \text{ and isProof}(l), \\ k \mathbin{\overline{\mp}} l & \text{otherwise.} \end{cases}$$

Note that the functions $\oplus$, h, and $\mp$ are all total recursive. It is also easy to see that each of the three conditions on k and l is a binary primitive recursive relation and that these relations are pairwise disjoint. Hence, by Remark 6.12, the function $\tilde{\oplus}$ is total recursive. For the same reasons, the following two functions are also total recursive:

$$x \mathbin{\tilde{\otimes}} y := \begin{cases} k \otimes l & \text{if isProof}(k) \text{ and isProof}(l), \\ k \otimes \mathsf{h}(l) & \text{if isProof}(k) \text{ and } \overline{\mathsf{isProof}}(l), \\ \mathsf{h}(k) \otimes l & \text{if } \overline{\mathsf{isProof}}(k) \text{ and isProof}(l), \\ k \mathbin{\overline{\times}} l & \text{otherwise;} \end{cases}$$

and

$$\tilde{\mathsf{e}}(k) := \begin{cases} \mathsf{g}(k) \otimes \mathsf{e}(k) & \text{if isProof}(k), \\ \overline{!}(k) & \text{otherwise.} \end{cases}$$

Lemma 6.50. $(\mathsf{Prf}, \tilde{\oplus}, \tilde{\otimes}, \tilde{\mathsf{e}})$ *is a normal proof predicate.*

Proof. Prf is a proof predicate by (6.27) and Lemmas 6.24 and 6.41. It remains to show the implications (6.30)–(6.32). Given the constant oscillation of reasoning between recursive functions on natural numbers and arithmetical formulas on numerals, we do not omit any underlinings here. Note that for the primitive recursive relation Proof we can freely move between equivalent statements $\mathsf{Proof}(m,n)$ and $\mathbb{N} \vDash \underline{\mathsf{Proof}}(\underline{m}, \underline{n})$, whereas for the provably Δ_1 formula $\mathsf{Prf}(x, y)$ only the latter can be formulated.

We start with (6.30). Suppose
$$\mathbb{N} \models \mathsf{Prf}(\underline{m}, \ulcorner\phi \to \psi\urcorner) \quad \text{and} \quad \mathbb{N} \models \mathsf{Prf}(\underline{n}, \ulcorner\phi\urcorner) \ .$$
There are four cases to consider, one for each case in the definition of $\tilde{\otimes}$.

1. $\mathsf{isProof}(m)$ and $\mathsf{isProof}(n)$, i.e., both m and n code 'real' proofs (according to $\mathsf{isProof}$). Then so does $m \,\tilde{\otimes}\, n$. Since m and n code sequences of (codes of) arithmetical formulas, they cannot be codes of evidence terms, hence, by (6.27), we have that
$$\mathbb{N} \models \underline{\mathsf{Proof}(\underline{m}, \ulcorner\phi \to \psi\urcorner)} \quad \text{and} \quad \mathbb{N} \models \underline{\mathsf{Proof}(\underline{n}, \ulcorner\phi\urcorner)} \ .$$
Thus, $\ulcorner\phi \to \psi\urcorner$ must occur in the sequence m and $\ulcorner\phi\urcorner$ must occur in the sequence n, it follows that $\ulcorner\psi\urcorner$ occurs in the sequence $m \otimes n$, i.e., $\mathsf{Proof}(m \otimes n, \ulcorner\psi\urcorner)$, and, hence, $\mathbb{N} \models \mathsf{Prf}(\underline{m \,\tilde{\otimes}\, n}, \ulcorner\psi\urcorner)$.

2. $\mathsf{isProof}(m)$ and $\overline{\mathsf{isProof}}(n)$. In this case, again
$$\mathbb{N} \models \underline{\mathsf{Proof}(\underline{m}, \ulcorner\phi \to \psi\urcorner)}$$
and, because of
$$\mathbb{N} \not\models \underline{\mathsf{Proof}(\underline{n}, \ulcorner\phi\urcorner)} \quad \text{and} \quad \mathbb{N} \models \mathsf{Prf}(\underline{n}, \ulcorner\phi\urcorner) \ ,$$
there exist an evidence term s with $n = \ulcorner s\urcorner$ and an $\mathcal{L}_\mathsf{J}$-formula F with $\phi = F^\dagger$ such that $\mathcal{M}_\mathsf{fin} \Vdash s{:}F$. Hence by the definition of the function h, we find $\mathsf{Proof}(\mathsf{h}(n), \ulcorner\phi\urcorner)$. We conclude that $\mathsf{Proof}(m \otimes \mathsf{h}(n), \ulcorner\psi\urcorner)$ and, hence,
$$\mathbb{N} \models \mathsf{Prf}(\underline{m \,\tilde{\otimes}\, n}, \ulcorner\psi\urcorner) \ .$$

3. $\overline{\mathsf{isProof}}(m)$ and $\mathsf{isProof}(n)$. This case is similar to the previous case.

4. In the remaining case, $\overline{\mathsf{isProof}}(m)$ and $\overline{\mathsf{isProof}}(n)$. Then there are evidence terms s and t and $\mathcal{L}_\mathsf{J}$-formulas F and G with
$$m = \ulcorner s\urcorner, \quad n = \ulcorner t\urcorner, \quad \phi = F^\dagger, \quad \text{and} \quad \psi = G^\dagger$$
such that
$$\mathcal{M}_\mathsf{fin} \Vdash s{:}(F \to G) \quad \text{and} \quad \mathcal{M}_\mathsf{fin} \Vdash t{:}F \ .$$
Thus, $\mathcal{M}_\mathsf{fin} \Vdash s{\cdot}t{:}G$. Since $\ulcorner s \cdot t\urcorner = m \,\overline{\times}\, n$, we conclude
$$\mathbb{N} \models \mathsf{Prf}(\underline{m \,\overline{\times}\, n}, \ulcorner G^\dagger\urcorner) \ ,$$
which is
$$\mathbb{N} \models \mathsf{Prf}(\underline{m \,\tilde{\otimes}\, n}, \ulcorner\psi\urcorner) \ .$$

6.4 Arithmetical Interpretation

Thus (6.30) is established.

We omit the proof of (6.31), which is similar to (6.30).

We now establish (6.32). Suppose $\mathbb{N} \models \mathsf{Prf}(\underline{n}, \ulcorner \phi \urcorner)$. There are two cases to consider.

1. $\mathsf{isProof}(n)$. In this case, we have that $\ulcorner \phi \urcorner$ occurs in the sequence n and $\mathbb{N} \models \underline{\mathsf{Proof}(\underline{n}, \ulcorner \phi \urcorner)}$. Hence, by the definition of the function e,

   $$\mathsf{Proof}\left(\mathsf{e}(n),\ \ulcorner \underline{\mathsf{Proof}(\underline{n}, \ulcorner \phi \urcorner)} \urcorner\right)$$

   and, by the definition of g,

   $$\mathsf{Proof}\left(\mathsf{g}(n),\ \ulcorner \underline{\mathsf{Proof}(\underline{n}, \ulcorner \phi \urcorner)} \to \mathsf{Prf}(\underline{n}, \ulcorner \phi \urcorner) \urcorner\right)\ .$$

   Clearly, now

   $$\mathsf{Proof}\left(\mathsf{g}(n) \otimes \mathsf{e}(n),\ \ulcorner \mathsf{Prf}(\underline{n}, \ulcorner \phi \urcorner) \urcorner\right)\ ,$$

   which yields

   $$\mathbb{N} \models \mathsf{Prf}\left(\underline{\mathsf{g}(n) \otimes \mathsf{e}(n)},\ \ulcorner \mathsf{Prf}(\underline{n}, \ulcorner \phi \urcorner) \urcorner\right)\ ,$$

   i.e.,

   $$\mathbb{N} \models \mathsf{Prf}\left(\underline{\tilde{\mathsf{e}}(n)},\ \ulcorner \mathsf{Prf}(\underline{n}, \ulcorner \phi \urcorner) \urcorner\right)\ .$$

2. In the remaining case $\overline{\mathsf{isProof}}(n)$. Then there exists an evidence term t and an $\mathcal{L}_\mathsf{J}$-formula F with $n = \ulcorner t \urcorner$ and $\phi = F^\dagger$ such that $\mathcal{M}_\mathsf{fin} \Vdash t{:}F$. Hence, $\mathcal{M}_\mathsf{fin} \Vdash !t{:}t{:}F$. Therefore,

   $$\mathbb{N} \models \mathsf{Prf}\left(\ulcorner !t \urcorner,\ \ulcorner (t{:}F)^\dagger \urcorner\right)\ .$$

   Since $\ulcorner !t \urcorner = \tilde{!}(\ulcorner t \urcorner) = \tilde{!}(n) = \tilde{\mathsf{e}}(n)$ and $(t{:}F)^\dagger = \mathsf{Prf}(\underline{\ulcorner t \urcorner}, \ulcorner F^\dagger \urcorner) = \mathsf{Prf}(\underline{n}, \ulcorner \phi \urcorner)$, we conclude

   $$\mathbb{N} \models \mathsf{Prf}\left(\underline{\tilde{\mathsf{e}}(n)},\ \ulcorner \mathsf{Prf}(\underline{n}, \ulcorner \phi \urcorner) \urcorner\right)\ ,$$

   which finishes the proof of (6.32). □

Lemma 6.51. *Let $*$ be a mapping such that $s^* := \ulcorner s \urcorner$ for each atomic evidence term s and $p^* := p^\dagger$ for each atomic proposition p. Then*

$$(*, \mathsf{Prf}, \tilde{\otimes}, \tilde{\oplus}, \tilde{\mathsf{e}})$$

is an arithmetical CS-interpretation.

Proof. $(\mathsf{Prf}, \tilde{\otimes}, \tilde{\oplus}, \tilde{\mathsf{e}})$ is a normal proof predicate by Lemma 6.50. Thus by the definition of $*$, we immediately get that $(*, \mathsf{Prf}, \tilde{\otimes}, \tilde{\oplus}, \tilde{\mathsf{e}})$ is an arithmetical interpretation.

It remains to show that this interpretation respects the constant specification CS. To establish this, we first show that for any evidence term t,
$$t^* = \ulcorner t \urcorner . \tag{6.37}$$
We proceed by induction on the structure of t and distinguish the following cases.

1. t is atomic. We have $t^* = \ulcorner t \urcorner$ by the definition of $*$.

2. $t = r \cdot s$. By the injectivity of our Gödel numbering, $\overline{\mathsf{isProof}}(\ulcorner r \urcorner)$ and $\overline{\mathsf{isProof}}(\ulcorner s \urcorner)$, so
$$(r \cdot s)^* = r^* \tilde{\otimes} s^* = \ulcorner r \urcorner \tilde{\otimes} \ulcorner s \urcorner = \ulcorner r \urcorner \overline{\times} \ulcorner s \urcorner = \ulcorner r \cdot s \urcorner .$$

3. $t = r + s$. By similar reasoning,
$$(r + s)^* = r^* \tilde{\oplus} s^* = \ulcorner r \urcorner \tilde{\oplus} \ulcorner s \urcorner = \ulcorner r \urcorner \overline{\mp} \ulcorner s \urcorner = \ulcorner r + s \urcorner .$$

4. $t = !s$. By similar reasoning,
$$(!s)^* = \tilde{\mathsf{e}}(s^*) = \tilde{\mathsf{e}}(\ulcorner s \urcorner) = \overline{!}(\ulcorner s \urcorner) = \ulcorner !s \urcorner .$$

Hence (6.37) is established. Now
$$F^* = F^\dagger \tag{6.38}$$
for any $\mathcal{L}_\mathsf{J}$-formula F can be shown as in the proof of Lemma 6.42. Using (6.38), we find that the constant specification CS is respected again as in the proof of Lemma 6.42. □

Theorem 6.52. *Let CS be a finite constant specification. For any formula F of $\mathcal{L}_\mathsf{J}$ we have*

$$F \text{ is arithmetically CS-valid} \quad \text{implies} \quad \mathsf{LP}_\mathsf{CS} \vdash F .$$

Proof. Assume that $\mathsf{LP}_\mathsf{CS} \nvdash F$. By Theorem 4.35, there exists a finitary CS-model $\mathcal{M}_\mathsf{fin}$ with $\mathcal{M}_\mathsf{fin} \nvDash F$. By Lemma 6.40, we have $\mathbb{N} \nvDash F^\dagger$. By Lemma 6.51, there is an arithmetical CS-interpretation $(*, \mathsf{Prf}, \tilde{\otimes}, \tilde{\oplus}, \tilde{\mathsf{e}})$ such that $F^* = F^\dagger$. Hence, $\mathbb{N} \nvDash F^*$. We have $\mathsf{PA} \nvdash F^*$ by the soundness of PA. We conclude that F is not arithmetically CS-valid. □

Remark 6.53. The choice of Peano Arithmetic for establishing arithmetical completeness of $\mathsf{LP}_{\mathsf{CS}}$ is somewhat arbitrary. We use it since it is a well-known theory of arithmetic and since it was used by Artemov [6] in the original proof of arithmetical completeness. However, Goris [75] shows that $\mathsf{LP}_{\mathsf{CS}}$ (for finite CS) is also arithmetically complete with respect to Buss's system S_2^1 of bounded arithmetic [39].

6.5 A Semantics of Proofs for Intuitionistic Logic

We claimed that the Logic of Proofs can be viewed as a formalization of the BHK semantics via the following chain of embeddings

$$\mathsf{IPC} \hookrightarrow \mathsf{S4} \hookrightarrow \mathsf{LP}_{\mathsf{CS}} \hookrightarrow \text{classical proofs} \ . \qquad (6.39)$$

There are two ways how to read (6.39). One is based on simple arithmetical interpretations, the other uses arithmetical interpretations. Let us start with the former.

Theorem 6.54. *Let CS be a primitive recursive, axiomatically appropriate, and schematic constant specification. There exists a realization r such that for each $\mathcal{L}_\square$-formula F*

$\mathsf{S4} \vdash F$ *if and only if* $r(F)$ *is weakly arithmetically CS-valid .*

Proof. First we show the direction from left to right. By Realization Theorem 3.22, there exists a realization r such that for each formula F of $\mathcal{L}_\square$

$\mathsf{S4} \vdash F$ implies $\mathsf{LP}_{\mathsf{CS}} \vdash r(F)$.

Combining this with Theorem 6.35 we obtain for each $\mathcal{L}_\square$-formula F

$\mathsf{S4} \vdash F$ implies $r(F)$ is weakly arithmetically CS-valid .

For the direction from right to left, let r be an arbitrary realization and suppose that $r(F)$ is weakly arithmetically CS-valid. By Theorem 6.37 we obtain $\mathsf{LP}_{\mathsf{CS}} \vdash r(F)$. Hence by Lemma 3.2 we find $\mathsf{S4} \vdash F$. □

To obtain a provability semantics for intuitionistic logic, we combine the previous result with the embedding of IPC into $\mathsf{S4}$. Recall from

Theorem 1.51 that the Gödel translation $\mathsf{Gt}(\cdot)$ from IPC to S4 is such that for each $\mathcal{L}_{\mathsf{ip}}$-formula F we have

$$\mathsf{IPC} \vdash F \quad \text{if and only if} \quad \mathsf{S4} \vdash \mathsf{Gt}(F) \ .$$

Corollary 6.55. *Let CS be a primitive recursive, axiomatically appropriate, and schematic constant specification. There exists a realization r such that for each $\mathcal{L}_{ip}$-formula F*

$$\mathsf{IPC} \vdash F$$

if and only if

$r(\mathsf{Gt}(F))$ *is weakly arithmetically CS-valid* .

Figure 6.1 illustrates how we obtain a provability interpretation for intuitionistic logic using the simple arithmetic semantics for $\mathsf{LP}_{\mathsf{CS}}$.

$$\mathsf{IPC} \vdash F \ \hookrightarrow_{\mathsf{Gt}} \ \mathsf{S4} \vdash \mathsf{Gt}(F) \ \longrightarrow^{r} \ \mathsf{LP}_{\mathsf{CS}} \vdash r(\mathsf{Gt}(F)) \ \succ_{*} \rightarrow \ \mathsf{PA}$$

Figure 6.1: Simple arithmetical semantics

Using arithmetical interpretations to read (6.39), we obtain the following result.

Theorem 6.56. *There exists a realization r such that for each $\mathcal{L}_\square$-formula F*

$$\mathsf{S4} \vdash F$$

if and only if

$r(F)$ *is arithmetically CS-valid*

for some finite constant specification CS .

Proof. From left to right. Let CS' be an axiomatically appropriate and schematic constant specification. By the realization theorem, there exists a realization r such that for each $\mathcal{L}_\square$-formula F

$$\mathsf{S4} \vdash F \quad \text{implies} \quad \mathsf{LP}_{\mathsf{CS}'} \vdash r(F) \ . \tag{6.40}$$

Hence, by Lemma 2.14 we find that for each $\mathcal{L}_\square$-formula F

$$\mathsf{S4} \vdash F \quad \text{implies} \quad \mathsf{LP}_{\mathsf{CS}} \vdash r(F) \text{ for some finite CS} \ . \tag{6.41}$$

6.5 A Semantics of Proofs for Intuitionistic Logic

By arithmetical soundness we conclude that for each $\mathcal{L}_\square$-formula F

$\mathsf{S4} \vdash F$

implies

$r(F)$ is arithmetically CS-valid

for some finite constant specification CS .

For the direction from right to left, let r be an arbitrary realization and suppose that $r(F)$ is arithmetically CS-valid for some finite CS. By Theorem 6.52 we obtain $\mathsf{LP}_\mathsf{CS} \vdash r(F)$. Hence by Lemma 3.2 we find $\mathsf{S4} \vdash F$. □

Finally, we obtain a provability semantics for intuitionistic logic based on arithmetical interpretations.

Corollary 6.57. *There exists a realization r such that for each $\mathcal{L}_{ip}$-formula F*

$\mathsf{IPC} \vdash F$

if and only if

$r(\mathsf{Gt}(F))$ *is arithmetically* CS*-valid*

for some finite constant specification CS .

Figure 6.2 illustrates this corollary. The constant specification CS' is axiomatically appropriate and schematic, see (6.40). CS_F and CS_G are finite constant specifications given by (6.41).

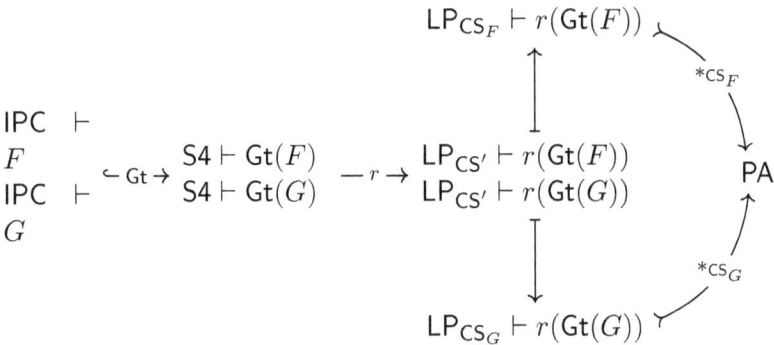

Figure 6.2: Arithmetical semantics

We see that in the case of arithmetical semantics, the provability interpretation depends on the formula we start with. In the case of simple arithmetical semantics, we can use the same interpretation for all formulas as shown in Figure 6.1).

6.6 Notes

The presentation and results of this chapter are based on papers by Artemov, Beklemishev, Goris, Kuznets, and Studer [4, 6, 11, 75, 101].

Artemov established arithmetical completeness and the provability semantics for intuitionistic logic (Corollary 6.57) already in the first papers on the Logic of Proofs [4, 6]. The notion of a simple arithmetical interpretation and the corresponding interpretation of intuitionistic logic (Corollary 6.55) have been introduced much later by Kuznets and Studer [101]. Note that they called it weak arithmetical interpretation.

We use the definition of a normal proof predicate from [75]. In Artemov's work a normal proof predicate is a proof predicate $\mathsf{Prf}(x,y)$ that satisfies the following two conditions:

1. *Finiteness of Proofs.* For any k, the set $T(k) := \{l \mid \mathsf{Prf}(k,l)\}$ is finite.

2. *Cojoinability of Proofs.* For any k and l, there is an n such that

$$T(k) \cup T(l) \subseteq T(n) \ .$$

These two conditions guarantee that given a proof predicate $\mathsf{Prf}(x,y)$, we can define functions $\tilde{\otimes}$, $\tilde{\oplus}$, and $\tilde{\mathsf{e}}$ such that $(\mathsf{Prf}(x,y), \tilde{\otimes}, \tilde{\oplus}, \tilde{\mathsf{e}})$ is a normal proof predicate in our sense.

Gödel [73] suggested using a system with explicit proofs for the interpretation of S4 already 1938, but that paper only appeared 1995. Even before the publication of Gödel's work, Artemov [4] came up with the Logic of Proofs $\mathsf{LP}_{\mathsf{CS}}$ and established a realization theorem as well as completeness with respect to arithmetical interpretations.

The first systems for logics of proofs that feature formulas of the form $t{:}F$, meaning *t is a proof of F*, appear in the work of Artemov and Strassen [20, 21, 22, 130] who investigate arithmetical interpretations for these logics. However, these ancestors of $\mathsf{LP}_{\mathsf{CS}}$ had no operations on proof terms and were too weak to capture the □-modality of S4 in full.

6.6 Notes

Artemov and Yavorskaya [23] present **FOLP**, a first-order variant of the Logic of Proofs that is capable of realizing first-order **S4** and, therefore, the first-order intuitionistic logic **HPC**. Since **FOLP** enjoys a natural provability interpretation, their work provides a semantics of explicit proofs for first-order **S4** and **HPC** compliant with Brouwer–Heyting–Kolmogorov requirements.

Tabatabai [133] provides a nice exposition of the arithmetical semantics for the Logic of Proofs. He also discusses some advantages and disadvantages of using the Logic of Proofs to give a BHK interpretation to intuitionistic logic.

Chapter 7

Self-Referentiality

Self-reference is a recurring topic in proof theory and epistemology. In the context of Peano Arithmetic, self-referring sentences received a thorough treatment in Smoryński [126]. It turns out that one can express many statements about self-reference in the language of modal logic where the notions of proof, Gödel number, and so on are abstracted away and hidden in the □-operator of Gödel–Löb logic.

In this chapter, we study self-referential proofs and justifications, i.e., valid proofs that prove statements about these same proofs. We show that such self-referential statements are present in the reasoning described by the modal logic S4. However, the modal language by itself is too coarse to capture this concept of self-referentiality. The language of justification logic will serve as an adequate refinement. Moreover, we establish that intuitionistic logic is self-referential in some strong sense, too.

7.1 Self-Referential Knowledge

A justification logics exhibits self-referentiality when a term t proves something about itself, that means

$$\mathsf{LP}_{\mathsf{CS}} \vdash t{:}F(t)$$

where $F(t)$ is a formula with at least one occurrence of the term t. Constructions of this kind are, of course, perfectly legal in the language of justification logic. In fact, when the constant specification CS is schematic, there must be self-referential theorems. Assume, for example, that a constant c justifies all instances of the axiom scheme jt, that

is all instances of $x{:}P \to P$. Then we have
$$\mathsf{LP_{CS}} \vdash c{:}(c{:}P \to P) \ ,$$
which is a self-referential theorem.

Let us illustrate how self-referentiality comes into play when realizing S4-theorems. Consider the following example, see also [13], which is inspired by the well-known Moore paradox:

> It will rain but I don't know that it will.

Example 7.1. If R stands for *it will rain*, then a modal formalization of the above sentence is
$$R \wedge \neg \Box R \ .$$
This formula is called the *Moore sentence*. It is easily satisfiable, e.g., when the weather forecast wrongly shows *no rain*. However, it is impossible to know the Moore sentence since
$$\neg \Box (R \wedge \neg \Box R) \tag{7.1}$$
is a theorem of S4.

A realization of (7.1) can be derived as follows. Let CS be an axiomatically appropriate and schematic constant specification. Obviously,
$$(R \wedge \neg x{:}R) \to R$$
is a tautology. Hence by constructive necessitation, there is a ground term t such that
$$\mathsf{LP_{CS}} \vdash t{:}((R \wedge \neg x{:}R) \to R) \ .$$
By Corollary 2.29, we can substitute $t \cdot x$ for x and get
$$\mathsf{LP_{CS}} \vdash t{:}((R \wedge \neg t \cdot x{:}R) \to R) \tag{7.2}$$
and thus
$$\mathsf{LP_{CS}} \vdash x{:}(R \wedge \neg t \cdot x{:}R) \to t \cdot x{:}R \ . \tag{7.3}$$
By the axiom scheme **jt** we have
$$\mathsf{LP_{CS}} \vdash x{:}(R \wedge \neg t \cdot x{:}R) \to (R \wedge \neg t \cdot x{:}R)$$
and thus
$$\mathsf{LP_{CS}} \vdash x{:}(R \wedge \neg t \cdot x{:}R) \to \neg t \cdot x{:}R \ . \tag{7.4}$$

7.1 Self-Referential Knowledge

From (7.3) and (7.4) we conclude by propositional reasoning

$$\mathsf{LP}_{\mathsf{CS}} \vdash \neg x{:}(R \wedge \neg t \cdot x{:}R) \ ,$$

which is a realization of (7.1). Note that the self-referential formula (7.2) is essential for this derivation.

Example 7.1 and Example 3.33 are two versions of the same idea. Indeed, the formula

$$\Box((P \to \Box P) \to \bot) \to \bot \qquad (3.11)$$

of Example 3.33 is logically equivalent to the formula expressing that the Moore sentence is not be known, cf. (7.1),

$$\neg\Box(P \wedge \neg\Box P) \ .$$

In the following we show that this use of self-referentiality was not just a bad choice we made but one that is inevitable. We establish that self-referentiality is inherent in the notion of knowledge axiomatized by S4 and $\mathsf{LP}_{\mathsf{CS}}$. That means we necessarily need a self-referential constant specification CS in order to realize S4 in $\mathsf{LP}_{\mathsf{CS}}$.

Definition 7.2 (Self-referential CS). A constant specification CS is called *directly self-referential* if $(c, A) \in \mathsf{CS}$ for some axiom A that contains at least one occurrence of the constant c.

Let $\mathsf{CS}_{\mathsf{nds}}$ be the maximal constant specification that is not directly self-referential, that is

$$\mathsf{CS}_{\mathsf{nds}} := \{(c, A) \mid c \text{ is a constant and } A \text{ is an axiom of LP such that}$$
$$c \text{ does not occur in } A\} \ .$$

In the next section we will show that self-referentiality is indeed inherent in S4. Our plan for the proof is as follows. We start with a particular formula G of $\mathcal{L}_\Box$ such that $\mathsf{S4} \vdash G$. Then we show that for any realization G^r of G, there is a generated $\mathsf{CS}_{\mathsf{nds}}$-model $\mathcal{M}$ such that $\mathcal{M} \not\Vdash G^r$. Hence we have $\mathsf{LP}_{\mathsf{CS}_{\mathsf{nds}}} \not\vdash G^r$. That means the S4-theorem G cannot be realized in $\mathsf{LP}_{\mathsf{CS}_{\mathsf{nds}}}$. Now since $\mathsf{CS}_{\mathsf{nds}}$ is the *maximal* constant specification that is not directly self-referential, we conclude that we necessarily need a direct self-referential constant specification CS in order to realize G in $\mathsf{LP}_{\mathsf{CS}}$.

We then use this result to establish that also intuitionistic logic is inherently self-referential. We start with a formula H of $\mathcal{L}_{ip}$ such that $\mathsf{CL} \vdash H$. Hence we have $\mathsf{IPC} \vdash \neg\neg H$ and $\mathsf{S4} \vdash \mathsf{Gt}(\neg\neg H)$ for the Gödel embedding $\mathsf{Gt}(\cdot)$ of IPC into CL. We show that $\mathsf{Gt}(\neg\neg H)$ can be used as the formula G above, which cannot be realized without self-referentiality. Hence we have a theorem of IPC such that any realization of its Gödel translation requires a direct self-referential constant specification.

7.2 Self-Referentiality of S4

Let F be a formula of $\mathcal{L}_\Box$ such that

$$\mathsf{S4} \not\vdash F \qquad (7.5)$$

and

$$F \text{ is an implication.} \qquad (7.6)$$

Consider now the $\mathcal{L}_\Box$-formula $\Diamond\Box F = \Box(\Box F \to \bot) \to \bot$. We abbreviate

$$G := \Box(\Box F \to \bot) \to \bot \ .$$

Any realization of G has the form

$$t_1{:}(t_2{:}F^r \to \bot) \to \bot$$

where F^r is a realization of F. We now fix such a realization of G, that is we fix t_1, t_2, and F^r.

We set

$$\mathcal{B}_0 := \mathsf{CS}_{\mathsf{nds}}$$
$$\mathcal{B} := \mathsf{CS}_{\mathsf{nds}} \cup \{(t_1, t_2{:}F^r \to \bot)\} \ .$$

Lemma 7.3. *For any subterm s of t_2 and any formula H of $\mathcal{L}_\mathsf{J}$:*

1. *If $(s, H) \in \mathcal{E}(\mathcal{B}_0)$, then*

    (a) $\mathsf{LP}_{\mathsf{CS}_{\mathsf{nds}}} \vdash H$ *and*

    (b) H *does not contain occurrences of t_2.*

2. *If $(s, H) \notin \mathcal{E}(\mathcal{B}_0)$ but $(s, H) \in \mathcal{E}(\mathcal{B})$, then*

    (a) H *has at least one occurrence of t_2,*

7.2 Self-Referentiality of S4

  (b) if H is an implication, then $H = t_2{:}F^r \to \bot$, and

  (c) $H \neq t_2{:}F^r$.

Proof. By induction on the structure of s. We distinguish the following cases.

$s = x$ is a variable.

1. We find $(x, H) \in \mathcal{B}_0 = \mathsf{CS}_{\mathsf{nds}}$, which is not possible. Thus Clause 1 is vacuously true.

2. We find $(x, H) \in \mathcal{B}$. Since $(x, H) \notin \mathcal{E}(\mathcal{B}_0) \supseteq \mathcal{B}_0$, we have
$$H = t_2{:}F^r \to \bot \ .$$
The latter does contain t_2 and is the only allowed implication.

$s = c$ is a constant.

1. We find $(c, H) \in \mathcal{B}_0 = \mathsf{CS}_{\mathsf{nds}}$. Thus H must be an axiom of LP and we have $\mathsf{LP}_{\mathsf{CS}_{\mathsf{nds}}} \vdash H$. Since $\mathsf{CS}_{\mathsf{nds}}$ is not directly self-referential, H cannot contain occurrences of c, which is a subterm of t_2. Thus H cannot contain t_2 either.

2. We find $(c, H) \in \mathcal{B}$. Since $(c, H) \notin \mathcal{E}(\mathcal{B}_0) \supseteq \mathcal{B}_0$, we have
$$H = t_2{:}F^r \to \bot \ .$$
The latter does contain t_2 and is the only allowed implication.

$s = s_1 + s_2$.

1. We find $(s_i, H) \in \mathcal{E}(\mathcal{B}_0)$ for $i = 1$ or $i = 2$. Thus by I.H. we have $\mathsf{LP}_{\mathsf{CS}_{\mathsf{nds}}} \vdash H$ and H does not contain occurrences of t_2.

2. There are two cases:

    (a) $t_1 = s_1 + s_2$ and $H = t_2{:}F^r \to \bot$. The latter does contain t_2 and is the only allowed implication.

    (b) $(s_i, H) \notin \mathcal{E}(\mathcal{B}_0)$ but $(s_i, H) \in \mathcal{E}(\mathcal{B})$ for $i = 1$ or $i = 2$. By I.H. we find that H contains t_2, if H is an impliciation, then $H = t_2{:}F^r \to \bot$, and $H \neq t_2{:}F^r$.

$s = s_1 \cdot s_2$.

1. In this case, there must exist a formula C such that

$$(s_1, C \to H) \in \mathcal{E}(\mathcal{B}_0) \text{ and } (s_2, C) \in \mathcal{E}(\mathcal{B}_0) \ .$$

By I.H. both $\mathsf{LP}_{\mathsf{CS}_{\mathsf{nds}}} \vdash C \to H$ and $\mathsf{LP}_{\mathsf{CS}_{\mathsf{nds}}} \vdash C$. Hence also $\mathsf{LP}_{\mathsf{CS}_{\mathsf{nds}}} \vdash H$. By I.H. we further know that $C \to H$ does not contain t_2, hence neither does H.

2. There are three cases:

   (a) $t_1 = s_1 \cdot s_2$ and $H = t_2{:}F^r \to \bot$. The latter does contain t_2 and is the only allowed implication.

   (b) There is a formula C with

   $$(s_1, C \to H) \in \mathcal{E}(\mathcal{B})$$
   $$(s_2, C) \in \mathcal{E}(\mathcal{B})$$
   $$(s_1, C \to H) \notin \mathcal{E}(\mathcal{B}_0) \ .$$

   We show that these three statements are, in fact, inconsistent. By I.H. for s_1 we find that

   $$C \to H \quad = \quad t_2{:}F^r \to \bot \ .$$

   So
   $$C = t_2{:}F^r \ ,$$

   Now we have two cases, both giving a contradiction.

   i. $(s_2, C) \notin \mathcal{E}(\mathcal{B}_0)$. We apply I.H. and obtain $C \neq t_2{:}F^r$. Contradiction.

   ii. $(s_2, C) \in \mathcal{E}(\mathcal{B}_0)$. We apply I.H. and obtain that C does not contain occurrences of t_2. Contradiction.

   (c) There is a formula C with

   $$(s_1, C \to H) \in \mathcal{E}(\mathcal{B})$$
   $$(s_2, C) \in \mathcal{E}(\mathcal{B})$$
   $$(s_2, C) \notin \mathcal{E}(\mathcal{B}_0) \ .$$

   We show that these three statements are also inconsistent. By I.H. for s_2 we find that C contains an occurrence of t_2. Thus the formula $C \to H$ also contains an occurrence of t_2. Now we have two cases, both giving a contradiction.

7.2 Self-Referentiality of S4

i. $(s_1, C \to H) \in \mathcal{E}(\mathcal{B}_0)$. We apply I.H. to s_1 and obtain that the formula $C \to H$ does not contain t_2. Contradiction.

ii. $(s_1, C \to H) \notin \mathcal{E}(\mathcal{B}_0)$. This is the same situation as in Case 2b.

$s = !s_1$.

1. We find $H = s_1{:}C$ for some formula C and $(s_1, C) \in \mathcal{E}(\mathcal{B}_0)$. Since s_1 is a subterm of t_2, we can apply I.H. and obtain that

$$C \text{ does not contain occurrences of } t_2 \qquad (7.7)$$

   and $\mathsf{LP}_{\mathsf{CS}_{\mathsf{nds}}} \vdash C$. Hence, by soundness (Theorem 4.12) we get $\mathcal{M} \Vdash C$ for any generated $\mathsf{CS}_{\mathsf{nds}}$-model $\mathcal{M}$. Because of $\mathcal{B}_0 = \mathsf{CS}_{\mathsf{nds}}$ we know by Lemma 4.7 that $(s_1, C) \in \mathcal{E}(\mathcal{C})$ for any $\mathcal{C} \supseteq \mathsf{CS}_{\mathsf{nds}}$. Therefore, $\mathcal{M} \Vdash s_1{:}C$ for any generated $\mathsf{CS}_{\mathsf{nds}}$-model $\mathcal{M}$. By completeness for generated CS-models (Theorem 4.16), we obtain $\mathsf{LP}_{\mathsf{CS}_{\mathsf{nds}}} \vdash s_1{:}C$, that is $\mathsf{LP}_{\mathsf{CS}_{\mathsf{nds}}} \vdash H$. Using (7.7) and that s_1 is a proper subterm of t_2, we conclude that H does not contain occurrences of t_2.

2. There are two cases:

   (a) $t_1 = !s_1$ and $H = t_2{:}F^r \to \bot$. The latter does contain t_2 and is the only allowed implication.

   (b) $H = s_1{:}C$ for some formula C with

$$(s_1, C) \in \mathcal{E}(\mathcal{B})$$
$$(s_1, C) \notin \mathcal{E}(\mathcal{B}_0) \ .$$

   By I.H. the formula C contains t_2, thus so does $H = s_1{:}C$. Since s_1 is a proper subterm of t_2, we find $H \neq t_2{:}F^r$. Last but not least, because H is not an implication, there is no additional claim to show and we are done. $\square$

Theorem 7.4. *For F, F^r, t_1, and t_2 as above we have*

$$\mathsf{LP}_{\mathsf{CS}_{\mathsf{nds}}} \nvdash t_1{:}(t_2{:}F^r \to \bot) \to \bot \ .$$

Proof. Consider $\mathcal{B}_0$ and $\mathcal{B}$ as given above. Assume

$$(t_2, F^r) \in \mathcal{E}(\mathcal{B}) \ . \qquad (7.8)$$

We have two cases.

1. $(t_2, F^r) \in \mathcal{E}(\mathcal{B}_0)$. By the previous lemma we find $\mathsf{LP}_{\mathsf{CS}_{\mathsf{nds}}} \vdash F^r$. This implies $\mathsf{S4} \vdash F$, which contradicts our assumption (7.5).

2. $(t_2, F^r) \notin \mathcal{E}(\mathcal{B}_0)$. By assumption (7.6) the outermost connective of F is an implication. Hence F^r is an implication, too. Therefore, by the previous lemma we conclude $F^r = t_2{:}F^r \to \bot$, which is not possible.

Since we have a contradiction in both cases, our assumption (7.8) must be false and we conclude

$$(t_2, F^r) \notin \mathcal{E}(\mathcal{B}) \ . \tag{7.9}$$

Consider the generated $\mathsf{CS}_{\mathsf{nds}}$-model $\mathcal{M} = (\mathsf{val}, \mathcal{B})$ where $\mathsf{val} = \mathsf{Prop}$. Because of (7.9) we have $\mathcal{M} \not\Vdash t_2{:}F^r$, which means $\mathcal{M} \Vdash t_2{:}F^r \to \bot$. By $(t_1, t_2{:}F^r \to \bot) \in \mathcal{B}$, we get $\mathcal{M} \Vdash t_1{:}(t_2{:}F^r \to \bot)$. Thus we obtain

$$\mathcal{M} \not\Vdash t_1{:}(t_2{:}F^r \to \bot) \to \bot \ .$$

Now $\mathsf{LP}_{\mathsf{CS}_{\mathsf{nds}}} \not\vdash t_1{:}(t_2{:}F^r \to \bot) \to \bot$ follows by soundness of $\mathsf{LP}_{\mathsf{CS}_{\mathsf{nds}}}$. □

Summing up, we have the following corollary.

Corollary 7.5. *Let F be a formula of $\mathcal{L}_\Box$ such that*

1. *F is an implication*

2. *$\mathsf{S4} \not\vdash F$.*

Then we have

$$\mathsf{LP}_{\mathsf{CS}_{\mathsf{nds}}} \not\vdash (\Diamond\Box F)^r$$

for any realization $(\Diamond\Box F)^r$ of $\Diamond\Box F$.

It remains to show the existence of a formula F such that $\mathsf{S4} \vdash \Diamond\Box F$ and F satisfies the assumptions of the above corollary.

Lemma 7.6. *Let F be the $\mathcal{L}_\Box$-formula $\Diamond\Box P \to P$. We have:*

1. *F is an implication*

2. *$\mathsf{S4} \not\vdash F$*

3. *$\mathsf{S4} \vdash \Diamond\Box F$.*

7.2 Self-Referentiality of S4

Proof. Obviously, F is an implication. To show

$$\mathsf{S4} \not\vdash F\ , \tag{7.10}$$

we give a countermodel. Let $W := \{1, 2\}$, $R := \{(1, 2),\ (1, 1),\ (2, 2)\}$, and $\mathsf{val}(P) = \{2\}$. Then $\mathcal{M} = (W, R, \mathsf{val})$ is a Kripke model for S4. We immediately find $\mathcal{M}, 1 \not\Vdash \Diamond\Box P \to P$ and (7.10) follows by soundness of S4.

To establish the third claim, we present an S4-proof of $\Diamond\Box F$. We first need some auxiliary derivations in S4.

(D1) $\mathsf{S4} \vdash A \vee B$ implies $\mathsf{S4} \vdash \Diamond A \vee \Box B$.
Indeed, $\mathsf{S4} \vdash A \vee B$ stands for $\mathsf{S4} \vdash \neg A \to B$. By **(Nec)** we get $\mathsf{S4} \vdash \Box(\neg A \to B)$. By the axiom scheme k and **(MP)** we obtain $\mathsf{S4} \vdash \Box\neg A \to \Box B$, which is $\mathsf{S4} \vdash \Diamond A \vee \Box B$.

(D2) $\mathsf{S4} \vdash \Diamond A \vee B$ implies $\mathsf{S4} \vdash \Diamond A \vee \Box B$.
Suppose $\mathsf{S4} \vdash \Diamond A \vee B$. By **D1** we get $\mathsf{S4} \vdash \Diamond\Diamond A \vee \Box B$. Now observe that $\mathsf{S4} \vdash \Diamond\Diamond A \to \Diamond A$ follows from axiom scheme 4 by propositional reasoning. Thus we conclude $\mathsf{S4} \vdash \Diamond A \vee \Box B$.

(D3) $\mathsf{S4} \vdash \Diamond A \vee A$ implies $\mathsf{S4} \vdash \Diamond A$.
Suppose $\mathsf{S4} \vdash \Diamond A \vee A$. From axiom scheme t and propositional reasoning we get $\mathsf{S4} \vdash A \to \Diamond A$. Now $\mathsf{S4} \vdash \Diamond A$ easily follows by propositional reasoning.

We can now establish $\mathsf{S4} \vdash \Diamond\Box(\Diamond\Box P \to P)$ as follows. By propositional reasoning we find

$$\mathsf{S4} \vdash (\Diamond\Box P \to P) \vee \neg P\ .$$

By **(D1)** we get

$$\mathsf{S4} \vdash \Box(\Diamond\Box P \to P) \vee \Diamond\neg P\ .$$

Again by **(D1)** we get

$$\mathsf{S4} \vdash \Diamond\Box(\Diamond\Box P \to P) \vee \Box\Diamond\neg P\ .$$

By propositional reasoning we obtain

$$\mathsf{S4} \vdash \Diamond\Box(\Diamond\Box P \to P) \vee (\Box\Diamond\neg P \vee P)\ ,$$

which is

$$\mathsf{S4} \vdash \Diamond\Box(\Diamond\Box P \to P) \vee (\Diamond\Box P \to P)\ .$$

By **(D2)** we get

$$\mathsf{S4} \vdash \Diamond\Box(\Diamond\Box P \to P) \lor \Box(\Diamond\Box P \to P) \ .$$

Finally, **(D3)** yields

$$\mathsf{S4} \vdash \Diamond\Box(\Diamond\Box P \to P) \ . \qquad \Box$$

Now we can state that a direct self-referential constant specification is essential in order to establish a realization theorem.

Corollary 7.7. *If* $\mathsf{LP_{CS}}$ *realizes the* $\mathsf{S4}$*-theorem* $G := \Diamond\Box(\Diamond\Box P \to P)$ *(that is* $\mathsf{LP_{CS}} \vdash G^r$ *for some realization* G^r *of* G*), then* CS *has to be directly self-referential.*

7.3 Self-Referentiality of Intuitionistic Logic

In the previous section, we have shown that the modal logic $\mathsf{S4}$ is inherently self-referential. Since intuitionistic logic IPC can be embedded into $\mathsf{S4}$, it is a natural question whether IPC also is self-referential. This is not a priori clear. It might be that performing a Gödel translation of an intuitionistic theorem always yields a modal formula that can be realized in $\mathsf{LP_{CS_{nds}}}$, that is without calling for a self-referential constant specification.

In this section we show that intuitionistic logic also is inherently self-referential. We present a large class of theorems of IPC such that in order to realize the Gödel translation of them, a self-referential constant specification is necessarily needed.

In the following $\mathsf{Gt}(\cdot)$ is the Gödel translation $\mathsf{Gt}(\cdot)$ from $\mathcal{L}_{\mathrm{ip}}$ to $\mathcal{L}_\Box$, which we have introduced in Definition 1.50.

Theorem 7.8. *Let* F *be a formula of* $\mathcal{L}_{ip}$ *such that*

1. *$\mathsf{IPC} \nvdash F$ but $\mathsf{CL} \vdash F$*

2. *F is an implication.*

Then we have

1. *$\mathsf{IPC} \vdash \neg\neg F$*

2. *$\mathsf{LP_{CS_{nds}}} \nvdash (\mathsf{Gt}(\neg\neg F))^r$ for any realization $(\mathsf{Gt}(\neg\neg F))^r$ of $\mathsf{Gt}(\neg\neg F)$.*

7.4 Notes

Proof. First we observe that $\mathsf{IPC} \vdash \neg\neg F$ immediately follows from
$$\mathsf{CL} \vdash F$$
by Glivenko's Theorem 1.49.

Now we show that the assumptions of Corollary 7.5 are satisfied. Because of $\mathsf{IPC} \nvdash F$, we have $\mathsf{S4} \nvdash \mathsf{Gt}(F)$ by Theorem 1.51. Moreover, $\mathsf{Gt}(F)$ is an implication since F is an implication and the translation $\mathsf{Gt}(\cdot)$ preserves this property. Hence we can apply Corollary 7.5 and obtain $\mathsf{LP}_{\mathsf{CS}_{\mathsf{nds}}} \nvdash (\Diamond\Box\mathsf{Gt}(F))^r$ for any realization $(\Diamond\Box\mathsf{Gt}(F))^r$ of $\Diamond\Box\mathsf{Gt}(F)$. We finish the proof by observing that
$$\mathsf{Gt}(\neg\neg F) = \Box(\Box\mathsf{Gt}(F) \to \bot) \to \bot = \Diamond\Box\mathsf{Gt}(F) \ . \qquad \Box$$

To finally establish the inherent self-referentiality of intuitionistic logic, we need an implication F of $\mathcal{L}_{\mathrm{ip}}$ such that $\mathsf{IPC} \nvdash F$ but $\mathsf{CL} \vdash F$. An obvious choice for this implication is the law of double negation
$$\neg\neg P \to P \ . \qquad\qquad \textbf{(LDN)}$$
This is a implication that is classically provable but not intuitionistically. Hence the double negation of (1.8) is provable in IPC but, by Theorem 7.8, its Gödel translation cannot be realized in $\mathsf{LP}_{\mathsf{CS}_{\mathsf{nds}}}$. Thus we necessarily need a self-referential constant specification CS to realize IPC in $\mathsf{LP}_{\mathsf{CS}}$.

Remark 7.9. There is one issue with our proof that intuitionistic logic is inherently self-referential. Namely, it relies on a particular embedding $\mathsf{Gt}(\cdot)$ of IPC into $\mathsf{S4}$. Now it could be that $\mathsf{Gt}(\cdot)$ is just a very bad choice for this embedding and that the observed self-referentiality is actually introduced by $\mathsf{Gt}(\cdot)$.

This, however, is not the case. Yu [147] was able to show that the self-referentiality result for intuitionistic logic is not sensitive to the choice of the embedding. He considered a general class of Gödel-style translations from IPC to $\mathsf{S4}$ and proved that for each translation of this class, there are IPC-theorems such that their translation cannot be realized in $\mathsf{LP}_{\mathsf{CS}}$ without using a direct self-referential constant specification CS.

7.4 Notes

The first self-referentiality result with respect to $\mathsf{S4}$ and $\mathsf{LP}_{\mathsf{CS}}$ has been obtained by Kuznets [32]. He considered the $\mathsf{S4}$-theorem
$$\Diamond(P \to \Box P) \ ,$$

which is logically equivalent to (7.1), and showed that it cannot be realized in $\mathsf{LP}_{\mathsf{CS}_{\mathsf{nds}}}$.

Later Kuznets [92, 93, 97] extended this result. He proved that also the modal logics K4, D4, and T cannot be realized in their corresponding justification counterparts without calling for a direct self-referential constant specification. However, not all modal logics are self-referential. Kuznets established that K and D are realizable in their justification counterparts for some non-self-referential constant specification.

The presentation of this chapter is based on [145] where Yu presents first self-referentiality results for intuitionistic logic. In a subsequent article [147], he introduces a very large class of Gödel-style embeddings and showed that his self-referentiality results hold for all embeddings of this class. This is a sign that self-referentiality is indeed inherent in intuitionistic logic and not an artifact of the embedding into S4.

Since Kuznets gave the first example of self-referentiality in S4, there was the question of a criterion for self-referentiality: is there a sufficient and necessary condition on S4-theorems for being realizable only with self-referential constant specifications? Yu [144] presents a necessary condition for self-referentiality: if an S4-theorem has a proof in G3s (a sequent calculus for S4, see [136]) without "prehistoric loops" (a certain form of impredicativity), then this theorem has a realization that does not call for a self-referential constant specification. Note that given an S4-theorem, it is decidable whether it has a prehistoric-loop-free proof [146]. Gass and Studer [62] showed that prehistoric loops are not sufficient for self-referentiality, but that with an expansion on the definition of self-referentiality, prehistoric loops become sufficient.

Bibliography

[1] Juan Pablo Aguilera and David Fernández-Duque. Verification logic: An arithmetical interpretation for negative introspection. In L. Beklemishev, S. Demri, and A. Máté, editors, *Advances in Modal Logic, Volume 11*, pages 1–20. College Publications, 2016.

[2] Jesse Alt and Sergei Artemov. Reflective λ-calculus. In Reinhard Kahle, Peter Schroeder-Heister, and Robert Stärk, editors, *Proof Theory in Computer Science, International Seminar, PTCS 2001, Dagstuhl Castle, Germany, October 7–12, 2001, Proceedings*, volume 2183 of *Lecture Notes in Computer Science*, pages 22–37. Springer, 2001.

[3] Evangelia Antonakos. Explicit generic common knowledge. In Sergei Artemov and Anil Nerode, editors, *Logical Foundations of Computer Science, International Symposium, LFCS 2013, San Diego, CA, USA, January 6–8, 2013, Proceedings*, volume 7734 of *Lecture Notes in Computer Science*, pages 16–28. Springer, 2013.

[4] Sergei N. Artemov. Operational modal logic. Technical Report MSI 95–29, Cornell University, December 1995.

[5] Sergei N. Artemov. Explicit provability: the intended semantics for intuitionistic and modal logic. Technical Report CFIS 98–10, Cornell University, September 1998.

[6] Sergei N. Artemov. Explicit provability and constructive semantics. *Bulletin of Symbolic Logic*, 7(1):1–36, March 2001.

[7] Sergei N. Artemov. Unified semantics for modality and λ-terms via proof polynomials. In Kees Vermeulen and Ann Copestake, editors, *Algebras, Diagrams and Decisions in Language, Logic and*

Computation, volume 144 of *CSLI Lecture Notes*, pages 89–118. CSLI Publications, Stanford, 2002.

[8] Sergei Artemov. Justified common knowledge. *Theoretical Computer Science*, 357(1–3):4–22, July 2006.

[9] Sergei Artemov. The logic of justification. *The Review of Symbolic Logic*, 1(4):477–513, December 2008.

[10] Sergei N. Artemov. The ontology of justifications in the logical setting. *Studia Logica*, 100(1–2):17–30, April 2012. Published online February 2012.

[11] Sergei N. Artemov and Lev D. Beklemishev. Provability logic. In D. M. Gabbay and F. Guenthner, editors, *Handbook of Philosophical Logic, 2nd Edition*, volume 13, pages 189–360. Springer, 2005.

[12] Sergei Artemov and Eduardo Bonelli. The intensional lambda calculus. In Sergei N. Artemov and Anil Nerode, editors, *Logical Foundations of Computer Science, International Symposium, LFCS 2007, New York, NY, USA, June 4–7, 2007, Proceedings*, volume 4514 of *Lecture Notes in Computer Science*, pages 12–25. Springer, 2007.

[13] Sergei Artemov and Melvin Fitting. Justification logic. In Edward N. Zalta, editor, *The Stanford Encyclopedia of Philosophy*. Fall 2012 edition, 2012.

[14] Sergei Artemov and Melvin Fitting. *Justification Logic: Reasoning with Reasons*. Cambridge University Press, 2019.

[15] Sergei Artemov and Rosalie Iemhoff. The basic intuitionistic logic of proofs. *Journal of Symbolic Logic*, 72(2):439–451, June 2007.

[16] Sergei Artemov and Roman Kuznets. Logical omniscience via proof complexity. In Zoltán Ésik, editor, *Computer Science Logic, 20th International Workshop, CSL 2006, 15th Annual Conference of the EACSL, Szeged, Hungary, September 25–29, 2006, Proceedings*, volume 4207 of *Lecture Notes in Computer Science*, pages 135–149. Springer, 2006.

[17] Sergei Artemov and Roman Kuznets. Logical omniscience as a computational complexity problem. In Aviad Heifetz, editor, *Theoretical Aspects of Rationality and Knowledge, Proceedings of the Twelfth Conference (TARK 2009)*, pages 14–23, Stanford University, California, July 6–8, 2009. ACM.

[18] Sergei Artemov and Roman Kuznets. Logical omniscience as infeasibility. *Annals of Pure and Applied Logic*, 165(1):6–25, January 2014. Published online August 2013.

[19] Sergei Artemov and Elena Nogina. Introducing justification into epistemic logic. *Journal of Logic and Computation*, 15(6):1059–1073, December 2005.

[20] Sergei Artëmov and Tyko Straßen. The basic logic of proofs. Technical Report iam–92–018, Institute of Computer Science and Applied Mathematics, University of Bern, 1992. Later version published as [21].

[21] Sergei Artëmov and Tyko Straßen. The basic logic of proofs. In E. Börger, G. Jäger, H. Kleine Büning, S. Martini, and M. M. Richter, editors, *Computer Science Logic, 6th Workshop, CSL'92, San Miniato, Italy, September 28–October 2, 1992, Selected Papers*, volume 702 of *Lecture Notes in Computer Science*, pages 14–28. Springer, 1993.

[22] Sergei Artëmov and Tyko Straßen. The logic of the Gödel proof predicate. In Georg Gottlob, Alexander Leitsch, and Daniele Mundici, editors, *Computational Logic and Proof Theory, Third Kurt Gödel Colloquium, KGC'93, Brno, Czech Republic, August 24–27, 1993, Proceedings*, volume 713 of *Lecture Notes in Computer Science*, pages 71–82. Springer, 1993.

[23] Sergei N. Artemov and Tatiana Yavorskaya (Sidon). First-order logic of proofs. Technical Report TR–2011005, CUNY Ph.D. Program in Computer Science, May 2011.

[24] Franz Baader and Wayne Snyder. Unification theory. In J.A. Robinson and A. Voronkov, editors, *Handbook of Automated Reasoning*, volume I, pages 447–533. Elsevier Science Publishers, 2001.

[25] Alexandru Baltag, Bryan Renne, and Sonja Smets. The logic of justified belief change, soft evidence and defeasible knowledge.

In Luke Ong and Ruy de Queiroz, editors, *Logic, Language, Information and Computation, 19th International Workshop, WoLLIC 2012, Buenos Aires, Argentina, September 3–6, 2012, Proceedings*, volume 7456 of *Lecture Notes in Computer Science*, pages 168–190. Springer, 2012.

[26] Alexandru Baltag, Bryan Renne, and Sonja Smets. The logic of justified belief, explicit knowledge, and conclusive evidence. *Annals of Pure and Applied Logic*, 165(1):49–81, January 2014. Published online in August 2013.

[27] Francisco Bavera and Eduardo Bonelli. Justification logic and history based computation. In Ana Cavalcanti, David Deharbe, Marie-Claude Gaudel, and Jim Woodcock, editors, *Theoretical Aspects of Computing – ICTAC 2010, 7th International Colloquium, Natal, Rio Grande do Norte, Brazil, September 1-3, 2010, Proceedings*, volume 6255 of *Lecture Notes in Computer Science*, pages 337–351. Springer, 2010.

[28] Patrick Blackburn, Maarten de Rijke, and Yde Venema. *Modal Logic*, volume 53 of *Cambridge Tracts in Theoretical Computer Science*. Cambridge University Press, 2002.

[29] Eduardo Bonelli and Federico Feller. Justification logic as a foundation for certifying mobile computation. *Ann. Pure Appl. Logic*, 163(7):935–950, 2012.

[30] George Boolos. *The Logic of Provability*. Cambridge University Press, 1993.

[31] Annemarie Borg and Roman Kuznets. Realization theorems for justification logics: Full modularity. In Hans De Nivelle, editor, *Automated Reasoning with Analytic Tableaux and Related Methods: 24th International Conference, TABLEAUX 2015, Wroclaw, Poland, September 21-24, 2015, Proceedings*, pages 221–236. Springer International Publishing, 2015.

[32] Vladimir Brezhnev and Roman Kuznets. Making knowledge explicit: How hard it is. *Theoretical Computer Science*, 357(1–3):23–34, July 2006.

[33] Samuel Bucheli, Meghdad Ghari, and Thomas Studer. Temporal justification logic. In S. Ghosh and R. Ramanujam, editors, *Proceedings of Methods for Modalities M4M-9*, volume 243 of *EPTCS*, pages 59–74. Open Publishing Association, 2017.

[34] Samuel Bucheli, Roman Kuznets, Bryan Renne, Joshua Sack, and Thomas Studer. Justified belief change. In Xabier Arrazola and María Ponte, editors, *LogKCA-10, Proceedings of the Second ILCLI International Workshop on Logic and Philosophy of Knowledge, Communication and Action*, pages 135–155. University of the Basque Country Press, 2010.

[35] Samuel Bucheli, Roman Kuznets, and Thomas Studer. Justifications for common knowledge. *Journal of Applied Non-Classical Logics*, 21(1):35–60, January–March 2011.

[36] Samuel Bucheli, Roman Kuznets, and Thomas Studer. Partial realization in dynamic justification logic. In Lev D. Beklemishev and Ruy de Queiroz, editors, *Logic, Language, Information and Computation, 18th International Workshop, WoLLIC 2011, Philadelphia, PA, USA, May 18–20, 2011, Proceedings*, volume 6642 of *Lecture Notes in Artificial Intelligence*, pages 35–51. Springer, 2011.

[37] Samuel Bucheli, Roman Kuznets, and Thomas Studer. Decidability for justification logics revisited. In Guram Bezhanishvili, Sebastian Löbner, Vincenzo Marra, and Frank Richter, editors, *Logic, Language, and Computation, 9th International Tbilisi Symposium on Logic, Language, and Computation, TbiLLC 2011, Kutaisi, Georgia, September 26-30, 2011, Revised Selected Papers*, volume 7758 of *Lecture Notes in Computer Science*, pages 166–181. Springer, 2013.

[38] Samuel Bucheli, Roman Kuznets, and Thomas Studer. Realizing public announcements by justifications. *Journal of Computer and System Sciences*, 80(6):1046–1066, 2014.

[39] Samuel R. Buss. *Bounded Arithmetic*. Bibliopolis, 1986.

[40] Samuel R. Buss and Roman Kuznets. Lower complexity bounds in justification logic. *Annals of Pure and Applied Logic*, 163(7):888–905, July 2012. Published online November 2011.

[41] Alexander Chagrov and Michael Zakharyaschev. *Modal Logic*, volume 35 of *Oxford Logic Guides*. Oxford University Press, 1997.

[42] Timothy Chappell. Plato on knowledge in the *Theaetetus*. In Edward N. Zalta, editor, *The Stanford Encyclopedia of Philosophy*. Metaphysics Research Lab, Stanford University, winter 2013 edition, 2013.

[43] Alonzo Church. *Introduction to Mathematical Logic*. Princeton University Press, 1956.

[44] Stephen A. Cook. The complexity of theorem-proving procedures. In *Proceedings of the Third Annual ACM Symposium on Theory of Computing*, STOC '71, pages 151–158. ACM, 1971.

[45] Jacques Corbin and Michel Bidoit. A rehabilitation of Robinson's unification algorithm. In R. E. A. Mason, editor, *Information Processing 83, Proceedings of the 9th World Computer Congress, Paris, France, September 19–23, 1983*, pages 909–914. North-Holland/IFIP, 1983.

[46] Evgenij Dashkov. Arithmetical completeness of the intuitionistic logic of proofs. *Journal of Logic and Computation*, 21(4):665–682, August 2011. Published online August 2009.

[47] W. Dean and H. Kurokawa. From the Knowability Paradox to the existence of proofs. *Synthese*, 176(2):177–225, September 2010. Published online May 2009.

[48] Walter Dean and Hidenori Kurokawa. The Paradox of the Knower revisited. *Annals of Pure and Applied Logic*, 165(1):199–224, January 2014. Published online in August 2013.

[49] Fred Dretske. Is knowledge closed under known entailment? the case against closure. In M. Steup and E. Sosa, editors, *Contemporary Debates in Epistemology*, pages 13—26. Malden MA: Blackwell, 2005.

[50] Ronald Fagin, Joseph Y. Halpern, Yoram Moses, and Moshe Y. Vardi. *Reasoning about Knowledge*. MIT Press, 1995.

[51] Tuan-Fang Fan and Churn-Jung Liau. A logic for reasoning about justified uncertain beliefs. In Qiang Yang and Michael Wooldridge, editors, *Proc. IJCAI 2015*, pages 2948–2954. AAAI Press, 2015.

[52] Michael J. Fischer and Richard E. Ladner. Propositional dynamic logic of regular programs. *Journal of Computing and System Science*, 18(2):194–211, 1979.

[53] Melvin Fitting. The logic of proofs, semantically. *Annals of Pure and Applied Logic*, 132(1):1–25, February 2005.

[54] Melvin Fitting. Realizations and LP. *Annals of Pure and Applied Logic*, 161(3):368–387, December 2009. Published online August 2009.

[55] Melvin Fitting. Justification logics and hybrid logics. *Journal of Applied Logic*, 8(4):356–370, December 2010. Published online August 2010.

[56] Melvin Fitting. The realization theorem for S5: A simple, constructive proof. In Johan van Benthem, Amitabha Gupta, and Eric Pacuit, editors, *Games, Norms and Reasons: Logic at the Crossroads*, volume 353 of *Synthese Library*, chapter 4, pages 61–76. Springer, 2011.

[57] Melvin Fitting. Realization using the model existence theorem. *Journal of Logic and Computation*, Advance Access, July 2013. Published online July 2013.

[58] Melvin Fitting. Justification logics and realization. Technical Report TR–2014004, CUNY Ph.D. Program in Computer Science, March 2014.

[59] Melvin Fitting. Possible world semantics for first-order logic of proofs. *Annals of Pure and Applied Logic*, 165(1):225–240, January 2014. Published online in August 2013.

[60] Dov M. Gabbay. *Labelled Deductive Systems*. Oxford University Press, 1996.

[61] James Garson. Modal logic. In Edward N. Zalta, editor, *The Stanford Encyclopedia of Philosophy*. Metaphysics Research Lab, Stanford University, summer 2014 edition, 2014. http://plato.stanford.edu/archives/sum2014/entries/logic-modal/.

[62] Nathan Sebastian Gass and Thomas Studer. Self-referentiality in justification logic. E-print 1902.01106, arXiv.org, 2019.

[63] Edmund L. Gettier. Is justified true belief knowledge? *Analysis*, 23(6):121–123, 1963.

[64] Meghdad Ghari. Distributed knowledge justification logics. *Theory of Computing Systems*, Online First, August 2013. Published online August 2013.

[65] Meghdad Ghari. Justification logics in a fuzzy setting. E-print 1407.4647, arXiv.org, 2014.

[66] Meghdad Ghari. Labeled sequent calculus for justification logics. *Annals of Pure and Applied Logic*, 2016.

[67] Alessandro Giordani. A logic of justification and truthmaking. *The Review of Symbolic Logic*, 6(2):323–342, June 2013. Published online March 2013.

[68] Jean-Yves Girard. Linear logic. *Theoretical Computer Science*, 50:1–102, 1987.

[69] Jean-Yves Girard. *Proof Theory and Logical Complexity*, volume 1 of *Studies in Proof Theory*. Bibliopolis, 1987.

[70] Valery Glivenko. Sur quelque points de la logique de M. Brouwer. *Academie Royale de Belgique. Bulletins de la Classe des Sciences, serie 5*, 15:183–188, 1929.

[71] Kurt Gödel. Über formal unentscheidbare Sätze der Principia Mathematica und verwandter Systeme I. *Monatshefte für Mathematik und Physik*, 38:173–198, 1931.

[72] Kurt Gödel. Eine Interpretation des intuitionistischen Aussagenkalküls. In *Ergebnisse eines Mathematischen Kolloquiums*, pages 39–40. 1933.

[73] Kurt Gödel. Vortrag bei Zilsel/Lecture at Zilsel's (*1938a). In Solomon Feferman, John W. Dawson, Jr., Warren Goldfarb, Charles Parsons, and Robert M. Solovay, editors, *Unpublished essays and lectures*, volume III of *Kurt Gödel Collected Works*, pages 86–113. Oxford University Press, 1995.

[74] Remo Goetschi and Roman Kuznets. Realization for justification logics via nested sequents: Modularity through embedding.

Annals of Pure and Applied Logic, 163(9):1271–1298, September 2012. Published online March 2012.

[75] Evan Goris. Feasible operations on proofs: The Logic of Proofs for bounded arithmetic. *Theory of Computing Systems*, 43(2):185–203, August 2008. Published online October 2007.

[76] Arend Heyting. Die intuitionistische Grundlegung der Mathematik. *Erkenntnis*, 2:106–115, 1931.

[77] Arend Heyting. *Mathematische Grundlagenforschung. Intuitionismus. Beweistheorie*. Springer, 1934.

[78] Jaakko Hintikka. *Knowledge and Belief: An Introduction to the Logic of the Two Notions*. Cornell University Press, 1962.

[79] George E. Hughes and Max J. Cresswell. *A New Introduction to Modal Logic*. Routledge, 1996.

[80] Giorgi Japaridze and Dick de Jongh. The logic of provability. In Samuel R. Buss, editor, *Handbook of Proof Theory*, volume 137 of *Studies in Logic and the Foundations of Mathematics*, chapter VII, pages 475–546. Elsevier, 1998.

[81] D. Kaplan and R. Montague. A paradox regained. *Notre Dame Journal of Formal Logic*, 1(3):79–90, 1960.

[82] Ioannis Kokkinis, Petar Maksimović, Zoran Ognjanović, and Thomas Studer. First steps towards probabilistic justification logic. *Logic Journal of IGPL*, 23(4):662–687, 2015.

[83] Ioannis Kokkinis, Zoran Ognjanović, and Thomas Studer. Probabilistic justification logic. In S. Artemov and A. Nerode, editors, *Proceedings of Logical Foundations of Computer Science LFCS'16*. Springer, 2016.

[84] A. Kolmogoroff. Zur Deutung der intuitionistischen Logik. *Mathematische Zeitschrift*, 35:58–65, 1932.

[85] Dénes König. Über eine Schlussweise aus dem Endlichen ins Unendliche. *Acta Litterarum ac Scientarum Ser. Sci. Math. Szeged*, 3:121–130, 1927.

[86] Vladimir N. Krupski. The single-conclusion proof logic and inference rules specification. *Annals of Pure and Applied Logic*, 113(1–3):181–206, December 2001.

[87] Vladimir N. Krupski. Reference constructions in the single-conclusion proof logic. *Journal of Logic and Computation*, 16(5):645–661, October 2006.

[88] Vladimir N. Krupski. Referential logic of proofs. *Theoretical Computer Science*, 357(1–3):143–166, July 2006.

[89] Hidenori Kurokawa. Tableaux and hypersequents for justification logics. *Annals of Pure and Applied Logic*, 163(7):831–853, July 2012. Published online October 2011.

[90] Roman Kuznets. On the complexity of explicit modal logics. In Peter G. Clote and Helmut Schwichtenberg, editors, *Computer Science Logic, 14th International Workshop, CSL 2000, Annual Conference of the EACSL, Fischbachau, Germany, August 21–26, 2000, Proceedings*, volume 1862 of *Lecture Notes in Computer Science*, pages 371–383. Springer, 2000. Errata concerning the explicit counterparts of $\mathcal{D}$ and $\mathcal{D}4$ are published as [94].

[91] Roman Kuznets. On decidability of the logic of proofs with arbitrary constant specifications. In *2004 Annual Meeting of the Association for Symbolic Logic, Carnegie Mellon University, Pittsburgh, PA, May 19–23, 2004*, volume 11(1) of *Bulletin of Symbolic Logic*, page 111. Association for Symbolic Logic, March 2005. Abstract.

[92] Roman Kuznets. *Complexity Issues in Justification Logic*. PhD thesis, City University of New York, May 2008.

[93] Roman Kuznets. Self-referentiality of justified knowledge. In Edward A. Hirsch, Alexander A. Razborov, Alexei Semenov, and Anatol Slissenko, editors, *Computer Science — Theory and Applications, Third International Computer Science Symposium in Russia, CSR 2008, Moscow, Russia, June 7–12, 2008, Proceedings*, volume 5010 of *Lecture Notes in Computer Science*, pages 228–239. Springer, 2008.

[94] Roman Kuznets. Complexity through tableaux in justification logic. In *2008 European Summer Meeting of the Association for*

Symbolic Logic, Logic Colloquium '08, Bern, Switzerland, July 3– July 8, 2008, volume 15(1) of *Bulletin of Symbolic Logic*, page 121. Association for Symbolic Logic, March 2009. Abstract.

[95] Roman Kuznets. A note on the use of sum in the Logic of Proofs. In Costas Drossos, Pavlos Peppas, and Constantine Tsinakis, editors, *Proceedings of the 7th Panhellenic Logic Symposium*, pages 99–103, Patras University, Greece, July 15–19, 2009. Patras University Press.

[96] Roman Kuznets. A note on the abnormality of realizations of S4LP. In Kai Brünnler and Thomas Studer, editors, *Proof, Computation, Complexity PCC 2010, International Workshop, Proceedings*, number IAM–10–001 in IAM Technical Reports. Institute of Computer Science and Applied Mathematics, University of Bern, June 2010. Abstract.

[97] Roman Kuznets. Self-referential justifications in epistemic logic. *Theory of Computing Systems*, 46(4):636–661, May 2010. Published online April 2009.

[98] Roman Kuznets, Sonia Marin, and Lutz Straßburger. Justification logic for constructive modal logic. IMLA 2017 - 7th Workshop on Intuitionistic Modal Logic and Applications, 2017.

[99] Roman Kuznets and Thomas Studer. Justifications, ontology, and conservativity. In Thomas Bolander, Torben Braüner, Silvio Ghilardi, and Lawrence Moss, editors, *Advances in Modal Logic, Volume 9*, pages 437–458. College Publications, 2012.

[100] Roman Kuznets and Thomas Studer. Update as evidence: Belief expansion. In Sergei Artemov and Anil Nerode, editors, *Logical Foundations of Computer Science, International Symposium, LFCS 2013, San Diego, CA, USA, January 6–8, 2013, Proceedings*, volume 7734 of *Lecture Notes in Computer Science*, pages 266–279. Springer, 2013.

[101] Roman Kuznets and Thomas Studer. Weak arithmetical interpretations for the logic of proofs. *Logic Journal of IGPL*, 24(3):424–440, 2016.

[102] Richard E. Ladner. The computational complexity of provability in systems of modal propositional logic. *SIAM Journal on Computing*, 6(3):467–480, 1977.

[103] Eveline Lehmann and Thomas Studer. Subset models for justification logic. E-print 1902.02707, arXiv.org, 2019.

[104] Leonid A. Levin. Universal search problems. *Problemy Peredachi Informatsii*, 9(3):115–115, 1973. Translated in [135].

[105] Michel Marti and Thomas Studer. Intuitionistic modal logic made explicit. *IfCoLog Journal of Logics and their Applications*, 3:877–901, 2016.

[106] Michel Marti and Thomas Studer. The internalized disjunction property for intuitionistic justification logic. In G. Bezhanishvili and G. D'Agostino, editors, *Advances in Modal Logic*, pages 1–20. College Publications, 2018.

[107] John-Jules Ch. Meyer and Wiebe van der Hoek. *Epistemic Logic for AI and Computer Science*. Cambridge University Press, 1995.

[108] J.C.C. McKinsey and Alfred Tarski. Some theorems about the sentential calculi of Lewis and Heyting. *The Journal of Symbolic Logic*, 13:1–15, 1948.

[109] Elliott Mendelson. *Introduction to Mathematical Logic*. Chapman & Hall, 5th edition, 2010.

[110] Robert Milnikel. Derivability in certain subsystems of the Logic of Proofs is Π_2^p-complete. *Annals of Pure and Applied Logic*, 145(3):223–239, March 2007.

[111] Robert S. Milnikel. The logic of uncertain justifications. *Annals of Pure and Applied Logic*, 165(1):305–315, January 2014. Published online in August 2013.

[112] Grigori Mints. Lewis' systems and system T (1965–1973). In Grigori Mints, editor, *Selected Papers in Proof Theory*, volume 3 of *Studies in Proof Theorie*, pages 221—294. Bibliopolis, 1992.

[113] Grigori Mints. *A Short Introduction to Intuitionistic Logic*. Kluwer, 2000.

[114] Alexey Mkrtychev. Models for the logic of proofs. In Sergei Adian and Anil Nerode, editors, *Logical Foundations of Computer Science, 4th International Symposium, LFCS'97, Yaroslavl, Russia, July 6–12, 1997, Proceedings*, volume 1234 of *Lecture Notes in Computer Science*, pages 266–275. Springer, 1997.

[115] Zoran Ognjanović, Nenad Savić, and Thomas Studer. Justification logic with approximate conditional probabilities. In A. Baltag, J. Seligman, and T. Yamada, editors, *6th International Conference on Logic, Rationality and Interaction, LORI VI*, volume 10455 of *LNCS*, pages 681–686. Springer, 2017.

[116] Nicholas Pischke. Dynamic extensions for the logic of knowing why with public announcements of formulas. E-print 1707.05617, arXiv.org, 2018.

[117] Konstantinos Pouliasis. A curry–howard view of basic justification logic. In Jouko Väänänen, Åsa Hirvonen, and Ruy de Queiroz, editors, *Logic, Language, Information, and Computation: 23rd International Workshop, WoLLIC 2016, Puebla, Mexico, August 16-19th, 2016. Proceedings*, pages 316–337. Springer Berlin Heidelberg, 2016.

[118] Konstantinos Pouliasis and Giuseppe Primiero. J-Calc: A typed lambda calculus for intuitionistic justification logic. In Valeria de Paiva, Mario Benevides, Vivek Nigam, and Elaine Pimentel, editors, *Proceedings of the 6th Workshop on Intuitionistic Modal Logic and Applications (IMLA 2013) in association with UNILOG 2013, Rio de Janeiro, Brazil, 7 April 2013*, number 300 in Electronic Notes in Theoretical Computer Science, pages 71–87. Elsevier, January 2014.

[119] Bryan Renne. Public communication in justification logic. *Journal of Logic and Computation*, 21(6):1005–1034, December 2011. Published online July 2010.

[120] Bryan Renne. Multi-agent justification logic: communication and evidence elimination. *Synthese*, 185(S1):43–82, April 2012. Published online July 2011.

[121] Barkley J. Rosser. Extensions of some theorems of Gödel and Church. *Journal of Symbolic Logic*, 1:87–91, 1936.

[122] Nenad Savić and Thomas Studer. Relevant justification logic. *Journal of Applied Logic*, in print.

[123] Walter J. Savitch. Relationships between nondeterministic and deterministic tape complexities. *Journal of Computer and System Sciences*, 4(2):177–192, 1970.

[124] A. Shen and N. K. Vereshchagin. *Computable Functions*, volume 19 of *Student Mathematical Library*. AMS, 2003. Translated from Russian by V. N. Dubrovskii.

[125] Stephen G. Simpson. *Subsystems of Second Order Arithmetic*. Springer, 1999.

[126] Craig Smoryński. *Self-Reference and Modal Logic*. Universitext. Springer-Verlag, 1985.

[127] Robert M. Solovay. Provability interpretations of modal logic. *Israel Journal of Mathematics*, 28:33–71, 1978.

[128] Richard Statman. Intuitionistic propositional logic is polynomial-space complete. *Theoretical Computer Science*, 9(1):67–72, 1979.

[129] Gabriela Steren and Eduardo Bonelli. The first-order hypothetical logic of proofs. *J. Log. Comput.*, 27(4):1023–1066, 2017.

[130] Tyko Straßen. *The Basic Logic of Proofs*. PhD thesis, Universität Bern, April 1994.

[131] Thomas Studer. Justified terminological reasoning. In E. Clarke, I. Virbitskaite, and A Voronkov, editors, *Proceedings of Perspectives of System Informatics PSI'11*, volume 7162 of *LNCS*, pages 349–361. Springer, 2012.

[132] Thomas Studer. Decidability for some justification logics with negative introspection. *Journal of Symbolic Logic*, 78(2):388–402, June 2013.

[133] Amirhossein Akbar Tabatabai. Provability interpretation of propositional and modal logics. E-print 1704.07677, arXiv.org, 2017.

[134] Alfred Tarski. A lattice-theoretical fixpoint theorem and its applications. *Pacific Journal of Mathematics*, 5:285–309, 1955.

[135] Boris A. Trakhtenbrot. A survey of russian approaches to perebor (brute-force searches) algorithms. *IEEE Ann. Hist. Comput.*, 6(4):384–400, 1984.

[136] Anne Sjerp Troelstra and Helmut Schwichtenberg. *Basic Proof Theory*. Cambridge University Press, second edition, 2000.

[137] Anne Sjerp Troelstra and Dirk van Dalen. *Constructivism in Mathematics, Vols. I and II*. North-Holland, 1988.

[138] Dirk van Dalen. Intuitionistic logic. In D. Gabbay and F. Guenthner, editors, *Handbook of Philosophical Logic, Vol. III: Alternatives to Classical Logic*, volume 166 of *Synthese Library*, chapter III.4, pages 225–339. D. Reidel Publishing Company, 1986.

[139] Hans van Ditmarsch, Wiebe van der Hoek, and Barteld Kooi. *Dynamic epistemic logic*. Springer, 2007.

[140] Chao Xu, Yanjing Wang, and Thomas Studer. A logic of knowing why. *Synthese*, in print.

[141] Tatiana Yavorskaya (Sidon). Logic of proofs and provability. *Annals of Pure and Applied Logic*, 113(1–3):345–372, December 2001.

[142] Tatiana Yavorskaya (Sidon). Interacting explicit evidence systems. *Theory of Computing Systems*, 43(2):272–293, August 2008. Published online October 2007.

[143] Rostislav E. Yavorsky. On arithmetical completeness of first-order logics of provability. In Frank Wolter, Heinrich Wansing, Maarten de Rijke, and Michael Zakharyaschev, editors, *Advances in Modal Logic, volume 3*, pages 1–16. World Scientific, 2002.

[144] Junhua Yu. Prehistoric phenomena and self-referentiality. In Farid Ablayev and Ernst W. Mayr, editors, *Computer Science — Theory and Applications, 5th International Computer Science Symposium in Russia, CSR 2010, Kazan, Russia, June 16–20, 2010, Proceedings*, volume 6072 of *Lecture Notes in Computer Science*, pages 384–396. Springer, 2010. Later version published as [146].

[145] Junhua Yu. Self-referentiality in the Brouwer–Heyting–Kolmogorov semantics of intuitionistic logic. In Sergei Artemov and Anil Nerode, editors, *Logical Foundations of Computer*

Science, International Symposium, LFCS 2013, San Diego, CA, USA, January 6–8, 2013, Proceedings, volume 7734 of *Lecture Notes in Computer Science*, pages 401–414. Springer, 2013. Later version published as [147].

[146] Junhua Yu. Prehistoric graph in modal derivations and self-referentiality. *Theory of Computing Systems*, 54(2):190–210, February 2014. Published online in December 2013.

[147] Junhua Yu. Self-referentiality of Brouwer–Heyting–Kolmogorov semantics. *Annals of Pure and Applied Logic*, 165(1):371–388, January 2014. Published online in August 2013.

Index

Π_2^p, 120, 127, 134
Σ_1-completeness of PA, 159
Σ_2^p, 120, 127, 134
$\oplus$, 178
$\otimes$, 178

almost schematic, 99, 112
(**AN**_{CS}), *see* axiom necessitation
annotation, 66
arithmetical competeness
    simple, 171
arithmetical completeness, 186
arithmetical interpretation, 180
    simple, 168
arithmetical soundness, 180
    simple, 169
Artemov, S., 57, 87, 111, 134, 137, 190
assignment, 147
ATm, *see* atomic Term
atomic terms, 30
axiom, 2
axiom necessitation, 33
axiom scheme, 3
axiom specification, 59
axiomatically appropriate, 33, 59

basic evaluation, 45
basic modular model, 46
Beklemishev, L., 190
BHK semantics, 135, 187, 190
bounded product function, 140

branch
    closed, 5
    open, 5
    satisfiable, 6
Brezhnev, V., 87
Brouwer, L. E. J., 135
Bucheli, S., 111
Buss, S., 134

canonical model for S4, 23
characteristic function, 142
CL, *see* classical propositional logic
$\mathsf{cl}_\mathcal{B}(X)$, 92
classical propositional language, 3
classical propositional logic, 4
closed branch, 5
closed tableau, 5
competeness
    simple arithmetical, 171
completed tableau, 7
completeness, 115
    of LP_{CS} w.r.t. basic modular models, 51
    of LP_{CS} w.r.t. finitary models, 110
    of LP_{CS} w.r.t. fully explanatory models, 55
    of LP_{CS} w.r.t. generated models, 96
    of LP_{CS} w.r.t. modular models, 54
    of S4, 26

tableau, 7
complexity
    of LP$_{CS}$, 121
    of a logic, 113
complexity class
    Π_2^p, 120, 127, 134
    Σ_2^p, 120, 127, 134
    coNP, 119, 126
    NP, 116, 126
    P, 115
    PSPACE, 119
computable function, 145
coNP, 119, 126
conservativity
    of LP$_{CS}$ over CL, 36
    of S4 over CL, 14
consistency of LP$_{CS}$, 48
consistent set of formulas, 20
constant, 30
constant specification, 32, 59
    almost schematic, 99, 112
    axiomatically appropriate, 33, 59
    directly self-referential, 195
    finite, 32, 59
    primitive recursive, 166
    schematic, 43, 59
    schematically injective, 134
    self-referential, 195
constructive necessitation, 40
constructive realization, 78, 81
Cook, S., 116
countable set, 2
CS, 32, 59
CS$_{nds}$, 195

decidability of LP$_{CS}$, 110
decidable, 90
deduction theorem, 37

deterministic polynomial problem, 115
deterministic Turing machine, 113
diagonalization lemma, 161
directly self-referential constant specification, 195
domain of a substitution, 71

e, 178
$\mathcal{E}(\mathcal{B})$, 93
embedding of
    CL in IPC, 27
    IPC in CL, 26
    IPC in S4, 27
empty sequence, 141
encoding of finite sequences, 140
equality of partial functions, 145
essential family, 81
evidence closure, 92
evidence relation, 93
evidence term, 30

factive evaluation, 46
factive quasimodel, 52
family of □-occurrences, 81
finitary model, 96
finite constant specification, 32, 59
finite model property, 90
Fitting, M., 58, 86, 87
forgetful projection, 62
formula
    provably Δ_1, 155
    provably Σ_1, 155
    signed, 5
    standard Σ_1, 155
    standard primitive recursive, 155
fully explanatory modular models, 55

Gabbay, D., 59

Gödel numbering, 160
Gödel's first Incompleteness theorem, 165
Gödel, K., 28, 136, 165, 190
Gödel–Löb logic, 137, 193
Gödel–Rosser incompleteness theorem, 163
generated model, 92, 93
Girard, J.-Y., 28
GL, *see* Gödel–Löb logic
Glivenko's theorem, 27
Goris, E., 190
ground term, 30
GS4, 79

hardness, 115
Heyting, A., 135

IPC, *see* intuitionistic propositional logic
incompleteness theorem, 163
induced model, 95
input variable, 71
internalization
    for arbitrary terms, 40
    for variables, 38
intuitionistic propositional language, 8
intuitionistic propositional logic, 9, 135, 195, 202
intuitionistic truth, 135

j, 31
j+, 31
j4, 31
JConst, *see* constant
jt, 31
justification language, 31
justification term, 30
JVar, *see* variable

Kolmogorov, A., 135
König's Lemma, 7
Kripke semantics, *see* possible world semantics
Kuznets, R., 87, 111, 134, 190, 203, 204

$\mathcal{L}_\Box$, *see* modal language
L-consistent set of formulas, 20
Ladner, R., 120
language of arithmetic, 146
law of double negation, 4, 203
$\mathcal{L}_{cp}$, *see* classical propositional language
LDN, *see* law of double negation
least fixed point, 92
Lehmann, E., 59
Levin, L., 116
lifting lemma, 78
Lindenbaum lemma, 20
$\mathcal{L}_{ip}$, *see* intuitionistic propositional language
live away from, 71
live on, 71
$\mathcal{L}_J$, *see* justification language
Logic of Proofs, 31, 33
LP, *see* Logic of Proofs
LP_0, 33
$\mathcal{L}_{PA}$, *see* language of arithmetic
LP_{CS}, 32

many-one reduction
    polynomial-time, 115
maximal consistent set of formulas, 20
McKinsey, J.C.C., 28, 136
mgu, *see* most general unifier
Milnikel, R., 59, 133, 134
Mkrtychev, A., 57, 111
modal language, 12

modal logic, 11
model existence, 50
modular model, 53
modus ponens, 5
monotonicity, 54
monotonicity of $\mathcal{E}$, 93
Moore sentence, 194
most general unifier, 98
(**MP**), *see* modus ponens
multiset, 79

$\mathbb{N}$, *see* natural number
natural number, 2
Nec, *see* necessitation rule
necessitation rule, 13
no new variable condition, 71
nondeterministic polynomial problem, 116
nondeterministic Turing machine, 113
normal proof predicate, 178
normal realization, 81
NP, 116, 126
numeral, 148

occurrence of a subformula, 65
open branch, 5
open tableau, 5
oracle computation, 120

P, 115
PA, *see* Peano Arithmetic
PAx1, **PAx2**, 4
PAx3–**PAx9**, 9
Peano Arithmetic, 149
polarity, 65
polarity of □, 80
polynomial hierarchy, 121
polynomial space problem, 119
polynomial-time many-one reduction, 115

possible world semantics
    for LP_CS, 51
    for S4, 17
Post's theorem, 90
potential pre-realization, 66
potential realization, 70
pre-realization
    potential, 66
primitive recursive function, 139
primitive recursive relation, 142
proof
    tableau, 6
proof predicate, 163
    normal, 178
propositional tableau, 5
provable equivalence, 155
provably Δ_1-formula, 155
provably Σ_1-formula, 155
PSPACE, 119

quasimodel, 51

rank of a term, 103
realization, 63, 138, 194
    constructive, 78, 81
    normal, 81
    potential, 70
    semantic, 77
    theorem, 77, 81
recursive function, 145
recursively enumerable, 90
relation, 141
replacement property, 44
restricted model, 109
$\mathsf{rk}(t)$, *see* rank of a term
Rosser, B., 165

S4, 13, 195
SAT, 116
satisfiability problem, 113

satisfiable branch, 6
Savitch, W., 119
schematic, 43, 59
schematically injective, 134
Schwichtenberg, H., 28
self-referential constant specification, 195
self-referentiality, 193
semantic realization, 77
sequent, 79
signed formula, 5
simple arithmetical competeness, 171
simple arithmetical interpretation, 168
simple arithmetical soundness, 169
soundness
   of $\mathsf{LP_{CS}}$ w.r.t. basic modular models, 46
   of $\mathsf{LP_{CS}}$ w.r.t. fully explanatory models, 55
   of $\mathsf{LP_{CS}}$ w.r.t. generated models, 94
   of $\mathsf{LP_{CS}}$ w.r.t. modular models, 54
   of $\mathsf{S4}$, 19
   simple arithmetical, 169
   tableau, 7
standard Σ_1-formula, 155
standard primitive recursive formula, 155
Statman, R., 120
Strassen, T., 190
Studer, T., 59, 111, 190
sub, 4, 13, 31
subformula
   in $\mathcal{L}_\square$, 13
   in $\mathcal{L}_{cp}$, 4
   in $\mathcal{L}_J$, 31

subset model, 59
substitution, 2, 41, 71
   domain of a, 71
   lives away from, 71
   lives on, 71
   no new variable condition, 71
   union, 71
substitution lemma, 42
substitution property, 43

Tabatabai, A., 191
tableau, 5
   closed, 5
   completed, 7
   open, 5
   propositional, 5
tableau completeness, 7
tableau construction rules, 6
tableau proof, 6
tableau soundness, 7
Tarski, A., 28, 136
term, *see* justification term
transitivity of p-reducibility, 116
Troelstra, A.S., 28
truth lemma, 50
Turing machine, 113

unifier, 98
union of multisets, 79
union of substitutions, 71

validity problem, 113
value of a term, 147
variable, 30
variable substitution, 71

weakly arithmetically CS-valid, 169

Yavorskaya, T., 190
Yu, J., 203, 204

www.ingramcontent.com/pod-product-compliance
Lightning Source LLC
Chambersburg PA
CBHW050139170426
43197CB00011B/1897